The Quantum Gamble

Jan C.A. Boeyens

The Quantum Gamble

Springer

Jan C.A. Boeyens
Centre for the Advancement of Scholarship
University of Pretoria
Pretoria, South Africa

ISBN 978-3-319-41620-5 ISBN 978-3-319-41621-2 (eBook)
DOI 10.1007/978-3-319-41621-2

Library of Congress Control Number: 2016950198

Printed on acid-free paper

This Springer imprint is published by Springer Nature
The registered company is Springer International Publishing AG Switzerland

Foreword

Quantum chemistry has developed to an extent that provides scientists with detailed information for understanding structures and properties of molecular compounds and assemblies, and software has been developed that allows to compute chemical systems and their properties in good agreement with experimental observations. Theoretical and computational chemistry has become an integral part of the education of scientists, and the computation of molecular and electronic structures, relative energies and pathways for transformations is an important tool for the interpretation of experimental data and the optimisation and even the 'rational design' of new systems in many areas of chemistry. Jan Boeyens' books discuss the fundamental basis of the physics of matter and quantum theory; their basis is a deep knowledge of the history of science, and they discuss features in quantum theory that are not much remembered and not directly included in modern computational chemistry—Jan's books and ideas are not mainstream; they are disputable and therefore are interesting and scientifically important. A computational model is not necessarily related to truth; its quality derives from the accuracy with which it simulates or predicts experimental observations. One of Jan Boeyens' concepts was to relate simplistic models as well as current computational chemistry approaches to basic notions of quantum theory. The present book summarises many of the results and hypotheses discussed in Jan Boeyens' earlier texts, and it presents these and new thoughts with as little mathematical formalism as possible: a textbook that is as general and generally understandable as possible and still fundamental, highly scientific and thorough, different from many other books in this area and therefore needs to be studied carefully, with a critical and open mind. This book was not planned to be Jan Boeyens' last book but unfortunately is—Jan died at the age of 81 just after finishing this manuscript.

Jan Boeyens was a prominent scientist in the area of theoretical and structural chemistry, very influential for the sciences in the African continent, especially South Africa, and beyond. Besides the careful analysis of hundreds of single-crystal structures, he has been deeply involved in method development for X-ray crystallography. In the early 1990s, he funded the 'Centre of Molecular Design' at the University of the Witwatersrand, a think tank and internationally visible school for computational chemistry. Jan Boeyens has also made enormous contributions to professional societies. Most of all, Jan Boeyens constantly asked important questions and was on the relentless search for their answers.

Heidelberg, Germany
June 2016

Peter Comba

Preface

More than a hundred years, ago a desperate manoeuvre by a German physicist resulted in a discovery that had the world of science in convulsion ever since. His name was Max Planck and he was of opinion that science was poised as a matter of course to find solutions to all but insignificant remnants of unsolved problems. His personal concern was to elucidate the conversion of heat into radiation, using the methods and laws of thermodynamics.

His initial subtle approach, based on appropriate mathematical models, produced nonsensical results such as the production of unlimited radiation energy from a modest input of heat. As an interim measure, he resorted to brute-force simulation to reproduce experimental data, with the intention that mature reflection would, in good time, eliminate the expedient but offensive enabling assumptions. This never happened.

Planck could never formulate a clear perception of the principle behind his successful simulation. The most puzzling aspect concerned the calculation of total energy as a discrete infinite sum, rather than by continuous integration, a procedure in conflict with the energy equipartition principle of thermodynamics.

Although Planck was not the only physicist who had a problem with the new discovery, he was one of the few who kept looking for a logical explanation. His more adventurous contemporaries recognised in it the emergence of a new nonclassical 'quantum' theory to replace classical physics.

The historical development and challenges of a new quantum paradigm during the early twentieth century are described in the first few chapters to follow.

The intransigence of, particularly, the particle lobby, in trying to obliterate the proposed wave model, led to the formulation of preposterous alternative concepts like probability waves, quantum uncertainty and subjective reality, developed into sensational philosophies by neophytes and eagerly endorsed by the masses. The difficulty of disputing this traditional consensus inspired by respect for authority rather than critical analysis is addressed. While many readers may be innocently unaware of the situation in quantum theory, some parallel, better publicised, scientific disputes are discussed for comparison.

The immediate objective was an understanding of spectroscopy and the nature of elementary matter. Both aspects are of immediate concern for chemistry and physics.

The major problem with quantum mechanics emerged in the theories of chemical matter, originating in a number of ill-conceived early assumptions and complicated by equally suspect subsequent embellishments. This is the big gamble that never paid off. The dilemma for chemistry arose from an uncritical acceptance of what appeared to be some harmless concepts from quantum physics. The most lethal are the assumption of wave-particle duality and the principle of linear superposition.

It is not only chemistry that suffered. As the perceived implications of quantum theory started to sink in, the philosophies of the Victorian era suffered an upheaval, over-reaction to which inaugurated a radically modified world view that persisted for the next century. This was a century of bitter dispute about issues, such as causality, non-Boolean logic, indeterminacy, free will, reductionism, non-locality and many other quasi-scientific concepts thought to be raised by the new physics. Only now has the pendulum slowly started to swing back. The time has come to reconsider the rigid interpretational models of twentieth-century physics and to eliminate obvious misconceptions that gained respectability through frequent uncritical repetition.

The new awareness behind the critical re-examination of quantum theory, in particular, stems from the recognition of nonlinearity as an important parameter in four-dimensional curved space-time. What complicates this exposition is that the implied theoretical modification is based on aspects of mathematics, which are largely ignored in modern curricula. A strictly non-mathematical account of essential concepts may hence become not only vague and unconvincing but, in some cases, practically impossible. Wherever words fail to communicate essential detail, mathematical formulation is provided in footnotes. On skipping these, unready readers, by accepting the conclusions on good faith, may hopefully be able to follow the main argument.

A sensible reassessment of quantum mechanics as a theory of chemistry for the uninitiated should start with a re-examination of the physical theories from which the new approach developed. This topic is addressed in Chap. 4.

A perceived anomaly of modern physics is the apparent inconsistency between its two major theories. Although the theories of relativity and the quantum developed in the hands of essentially the same group of scientists, a century's worth of refinement has failed to establish their common foundation. Quantum theory, which was the first to emerge as a nonclassical model of spectroscopic phenomena, soon developed into a search for an atomic model. The theory of special relativity, which established common ground between mechanical and electromagnetic motion, clearly addressed different aspects of the same phenomena that spawned the quantum theory. The conclusions reached by the two theories, however, could not be reconciled. The main disagreement centred around the speed which mechanical objects can attain and which the theory of relativity limits not to exceed the speed of light, whereas quantum phenomena appeared to be non-locally correlated. This philosophical dispute centred around the perceived probabilistic nature of quantum matter.

Schrödinger's formulation of a quantum-wave equation that could account for the electronic energy levels in the hydrogen atom and explain atomic spectra without the unphysical idea of quantum jumps never gained acceptance over the orthodoxy of dimensionless subatomic point particles. The same fate befell Madelung's interpretation of quantum effects in hydrodynamic analogy, until the ideas of quantum waves were revived by Bohm.

Experimental study of electric and magnetic phenomena during the nineteenth century showed that the spread of such interactions depends characteristically on the nature of the conducting medium, including the vacuum. The product of electric permittivity and magnetic permeability in the vacuum was found to have the dimensions of an inverse velocity squared, $\epsilon_0\mu_0 = 1/c^2$. The functional relationship between electric and magnetic fields was formulated by Maxwell as a wave equation, according to which a combined electromagnetic disturbance, such as visible light, propagates through the vacuum at the same constant velocity, c. The spread of a disturbance, once initiated in the vacuum, depends exclusively on ϵ_0 and μ_0 and carries no momentum imparted by the source. The inference that the motion of a light source has no effect on c appears to contradict the time-honoured notion of relative velocity as dictated by classical mechanics.

In order to avoid the discrepancy between relative motion in mechanical and electromagnetic systems, it is necessary to modify the classical formulation of relative motion in a given direction into a transformation that also affects the measurement of time. According to this transformation law, space and time coordinates cannot be specified independently. The transformation has the same form as an orthogonal rotation through an imaginary angle. As this rotation mathematically interchanges time and space coordinates, it means that they are symmetry related and no longer separable in the usual way. It is therefore more appropriate to deal with four-dimensional space-time, rather than the traditional three-dimensional space and absolute time.

On rotation in four dimensions, the familiar vector $\boldsymbol{r}$, which remains invariant of length $r = \sqrt{x^2 + y^2 + z^2}$ in three-dimensional rotation, now appears, in projection, to contract as it moves into the fourth dimension. This so-called Fitzgerald contraction is compensated for by dilation in the fourth, i.e. time and dimension. For objects that travel at a speed of $v << c$, the relativistic formulae reduce to the familiar expressions of classical mechanics. Sensible physics requires the two theories to be compatible, but the link between quantum theory and the electromagnetic field has not been identified covariantly.

Special relativity excludes accelerated motion and the gravitational field. This defect led Einstein to the formulation of a theory of general relativity. The seminal argument depends on a simple thought experiment which relies on the result of special relativity that a body in high-speed linear motion contracts but only in the direction of motion. The circumferential length of a circular object accelerated in orbital motion therefore contracts, but the radius remains unaffected.

The only explanation of this result is that acceleration distorts the geometry that fixes the value of π. It means that the geometry of space-time is no longer Euclidean but positively curved. The experimental demonstration that inertial mass (linked to the force generated by acceleration) and gravitational mass (linked to gravitational force) are equivalent resulted in the postulate of general relativity that matter is responsible for the curvature of space, which manifests as the gravitational field.

This action is modelled mathematically by Einstein's gravitational field equations. The important feature to note is the reciprocity between a curvature tensor and an energy-stress tensor. Curvature vanishes in Euclidean space. This observation confirms that space-time must be curved in order for matter to exist.

The development of physics is intimately connected with mathematics, in both content and structure. Too many quantum chemists appear to be blissfully unaware of the way in which physical models are strictly constrained by underlying mathematical number systems. This topic is briefly introduced and the relationship between major algebras, their characteristics and the link with linear number theory is outlined. The golden ratio of number theory is recognised as the single most important characteristic of space-time curvature, which is responsible for the ubiquity of self-similar spiral structures, from atoms to galaxies, and the periodicity of matter that underpins all of chemistry.

The geometrical requirement that any pair of straight lines intersect in a single point is at variance with the Euclidean axiom of parallel lines. The problem is overcome by adding an extra point at infinity that also contains negative infinity. The result is a straight line that turns back onto itself to define an involution and projective topology of space-time.

In Chaps. 6 and 7, the methods, results and abuse of traditional quantum theory and its relationship to the theory of relativity are examined more closely. It is a far cry from general relativity that predicts the appearance of matter in curved space-time and the rival theories of wave and particle quantum mechanics, formulated to describe atomic matter in terms of linear equations. The gradual progress towards a reconciliation of the theories has been characterised by misunderstandings, personal hostilities, professional rivalry and downright bigotry. Maybe time has dulled the original bitterness sufficiently by now for a calm reassessment of the situation.

The non-local action ascribed to wave-function collapse represents the major discrepancy between quantum and relativity theories. According to quantum theory, the components of a spin vector are linearly independent and of random orientation until measured in a magnetic field. A previously correlated pair of spins are assumed to remain governed by a single wave function as they drift apart. Whenever a measurement creates a spin component for one individual, the collapsing wave function instantaneously creates a correlated spin component for the second remote partner. Such non-local communication is prohibited by the theory of relativity.

The fallacy of this argument is exposed on noting the four-dimensional nature of spin. Not only is it described by the same rotation that represents the Lorentz transformation of relativity, but it constitutes a variable with four inseparable quaternionic components. Spin remains well defined in all directions and does not depend on any measurement. A correlated pair remains correlated, and whatever is measured on one side must be automatically valid on the remote site.

The final chapters get to grips with the objective to purge physical theory of mystical reasoning and outdated misconceptions in searching for a new paradigm based on appropriate mathematics, sound physics and sober philosophy. The major obstacle is the modern theory of particle physics, known as quantum chromodynamics (QCD). It is the most refined form of Newton's theory of matter, as composed of point particles which interact through the mediation of 'something else, which is not material.'

The particles of QCD are identified as unobservable objects, called quarks, with fractional electric charges, held together within mesons and nucleons by more unobservable agents, with well-defined properties, called gluons. The symmetry model that avoids all exclusion complications requires a total of 61 quarks and gauge bosons (gluons) with quantum numbers characteristic of isospin, strangeness, charm and colour eigenvalues. Mass is normalised empirically.

As an alternative working model, chiral matter is assumed to occur as wavelike nonlinear distortion of the four-dimensional aether in the double cover of closed projective space-time. The antipodal involuted domain is occupied by antimatter in dynamic equilibrium across the interface.

Electromagnetism is confined to the achiral interface in the form of bosonic waves, unimpeded by aether drag as recognised by special relativity. Neutrinos and cosmic rays that pervade the cosmos as pseudo-relativistic waves provide experimental evidence of residual elementary matter not incorporated into ponderable matter.

The four-dimensional harmonic modes of elementary waves define characteristic values of spin, charge and mass as observed for bosons, neutrinos, electrons and protons.

Confined to a limited region of space-time, the number of nodes of the resulting standing waves is traditionally interpreted as a quantum number. This is the number that baffled Planck and later interpreted in many fanciful ways.

In a highly nonlinear environment, the internodal segments separate into individual solitons, identified by Einstein as photons and as point particles by particle physicists.

The book is aimed at a general readership with high-school knowledge of science and especially also at the students and teachers of undergraduate chemistry and related sciences as a red flag. As a cautionary reminder, science should be practised with a critical open mind and with scant respect for authority. What the chemistry student has to glean from quantum theory is that elementary units of matter and radiation are best described as standing waves with intrinsic characteristic properties of spin, charge and mass in line with the harmonic modes of the wave. All of chemistry can be understood as the interaction between waves.

Pretoria, South Africa — Jan C.A. Boeyens

Contents

Chapter 1
The Mysterious Quantum

Twentieth-century physics was dominated by two theories which came into existence at the turn of the century. The two theories, known as *quantum mechanics* and *relativistic mechanics* respectively, developed as refinements of Newtonian mechanics, the reigning paradigm for three centuries, to take into account the electric and magnetic phenomena discovered during the previous century and succinctly explained as a single *electromagnetic field*.

The immediate objective to unify the new field with Newton's gravitational field was frustrated by a fundamental difference between the two fields. Whereas the gravitational field was interpreted in terms of particle interactions dictated by mechanical forces, electromagnetism was better described by field equations characteristic of wave motion.

A promising line of attack appeared to be through the medium of thermodynamics, considered to apply equally well to both systems. Theoretical efforts were concentrated towards a central problem which became known under the misnomer *black-body radiation*. It refers to the conversion of heat energy into radiation, as conveniently observed in the changing colour of a glowing block of metal, as a function of temperature. Theoretical models were constructed on the assumption of oscillators that respond to increased temperature by radiating at a different wavelength or frequency.

According to the Maxwell–Boltzmann distribution formula, applied to a glowing gas in a closed cavity (the black body), it was possible to integrate over the energies of an infinite number of oscillators in order to derive the mean vibrational energy at a given temperature and the associated wavelength of radiation. The predicted result, in complete accord with the fundamental *equipartition principle*, turned out to be disastrously wrong. Instead of predicting an optimal wavelength at a given temperature the calculated energy of radiation increased monotonically to infinity with decreasing wavelength.

This behaviour was ascribed by Max Planck, who first managed to solve the problem, to two related factors: a functional relationship between vibrational energy and the frequency of emitted radiation on the one hand, which dictates that the

J.C.A. Boeyens, *The Quantum Gamble*, DOI 10.1007/978-3-319-41621-2_1

integration over continuously changing vibrational states should be replaced by a discrete infinite sum. These modifications made all the difference and predicted the correct black-body spectrum, as observed experimentally, but in blatant conflict with classical thermodynamics. The success was due to the introduction of an additional parameter, known as Planck's constant, which defines a linear relationship between energy and frequency, at variance with the assumed equipartition of energy.

Planck himself, considered his calculation as an interim *ad hoc* procedure and refrained from trying to explain the meaning of the new universal constant, eventually named after him. More adventurous scientists who noticed that the constant relates an amount of energy, considered to be a particle property, to frequency, which is a wave property, soon came up with a number of innovative interpretations. The idea of contradicting the laws of thermodynamics was apparently too traumatic for Planck, who kept on searching for a 'reasonable' explanation. The most persuasive interpretation was proposed by Einstein, who obtained an alternative derivation of Planck's formula from the probability of transition between atomic levels that differ in energy by a discrete amount corresponding to the product of Planck's constant and the frequency of vibration.

Apart from the black-body problem, two other spectroscopic puzzles from the nineteenth century remained unresolved—the Balmer formula of atomic spectroscopy and the *photoelectric effect*. While the Balmer formula, considered no more than a numerological oddity, remained on the back burner, the photoelectric problem demanded immediate serious attention.

The seminal experimental observation was the emission of electrons from the surface of certain active metals under ultraviolet (uv) irradiation. Like visible light, uv light is a form of electromagnetic radiation, but of shorter invisible wavelength, and hence of higher energy. The puzzle was that each metal was characterized by a specific threshold frequency such that radiation of wavelength higher than the threshold was completely inactive towards that metal, irrespective of light intensity.

At frequencies above threshold the energy of emitted electricity, as measured in a retarding electric field, only depends on the frequency of the activating radiation. The total amount of emitted photoelectricity varies with the light intensity.

In view of the absolute consensus that electricity consists of a swarm of negatively-charged particles, known as *electrons*, the photoelectric effect becomes hard to understand, given the wave nature of electromagnetic radiation.

By proposing that the energy of a light beam consists of discrete packets or *quanta* of energy, proportional in magnitude to the frequency of the light, Einstein provided convincing explanations of black-body radiation and the photoelectric effect at the same time. It is readily visualized how a light quantum on striking an electron in the metal surface transfers the entire quantum of energy to the electron. For light of frequency higher than threshold the energy transferred to the electron is sufficient to overcome the attraction that confines it to the metal surface and it escapes with the excess as kinetic energy. A neat energy balance consistent with all observations could be demonstrated.

The *quantum theory* was born, characterized from the outset by an androgynous nature, which later became commonplace as *wave-particle duality*, without ever acquiring a convincing physical explanation. Light quanta became known as *photons*, each of energy = $h\times$frequency, where h is the well-known symbol for Planck's constant. This relationship soon became generalized by referring to all forms of energy as *quantized* in units of $h\nu$, ν = frequency. This dubious model has now survived for a century.

The radiation emitted by incandescent monatomic gases, such as hydrogen in electric discharge, presented another, related, scientific puzzle. Compared to ordinary light, which separates into a continuous colour spectrum on passage through a prism, the hydrogen spectrum consists of a number of emissions at sharply defined wavelengths, known as a line spectrum. A numerical relationship between the wavelengths of these emission lines, expressed in a simple formula involving a series of consecutive integers, was discovered by a school teacher named Johann Balmer.

An atomic model consistent with this observation and based on the planetary rings of the planet Saturn was proposed in 1904 by Hantaro Nagaoka [1]. He likened the rings to standing electric waves that surround an atomic nucleus, represented here by the planet, and showed that the electronic energies could be correlated with the Balmer formula.

The nuclear model of the atom was 'discovered' ten years later by Rutherford with only oblique reference to Nagaoka. The idea of a planetary atomic model and the simulation of Balmer's formula followed a few years later, proposed by Bohr, without any acknowledgement. Another decade elapsed before the idea of standing-wave stabilization of electronic energy was revived by de Broglie, again without recognition.

It is fair to say that none of Nagaoka's imitators ever grasped the full significance of his proposal. They all tried to re-interpret the concept as a particle or *quantum* model and the proposal remained dormant for a hundred years.

Nagaoka was, in all probability, not aware of Planck's black-body simulation and his work predated Einstein's photoelectric postulate. He accounted for Balmer's digital formula and atomic stability in terms of the integers associated with the wavelength of standing waves. Most likely for this reason, the quantum theorists of the day ignored his contribution, soon to be forgotten.

In retrospect the saturnian model is still superior to all main-stream nuclear atomic models, despite the fact that it is formulated in only two dimensions. It does not have the Bohr–Sommerfeld problem of radiative decay, which in reality had not been solved by standard Schrödinger[1] wave mechanics, and avoids the awkward pilot-wave construct of de Broglie.

The origin and nature of matter has been in philosophical dispute since time immemorial. The central question is to decide whether reality is continuous or

[1]In fairness it should be pointed out that if Schrödinger had his way the 3D saturnian model would have emerged from his work.

discrete and concerns the nature of the ultimate underlying substratum. In the early cosmology of Anaximander it appears as the eternal and infinite (boundless) *apeiron* that generated the opposites, hot–cold, wet–dry, *etc.*[2]

The Pythagoreans considered this to be *number*, while Leucippus and Democritus took the opposite view by postulating *atoms* as the primary form of matter, which cannot be further subdivided and which freely moves in a void. The void itself has no properties apart from its existence, its infinite extent, and its eternal unchanging nature.

Most of the authentic ancient sources have been lost, but a reliable account of classical atomic theory has survived as a poem by Lucretius [2]. It defines all matter to exist in the form of invisible and indestructable atoms in empty space. The properties of objects depend on the infinite variety and shapes of atoms. Solid matter results from close union between atoms by the entanglement of their interlocking shapes.[3]

The alternative view that held sway for more than a thousand years was formulated by Aristotle as a refinement of Anaximander's cosmology. The four elements: earth, water, air and fire, were assumed to constitute one continuous substance rather than a swarm of discrete particles in a void, which he rejected as a place empty of even any divine presence. In his view different substances resulted from the transmutation of the four-element primary matter into different forms and shapes. This view, which inspired alchemy, was supported by Descartes who held that matter, like the four elements, was infinitely divisible and massless.

A curious confused dichotomy appears in the writings of Newton. From religious conviction he accepted the views of Aristotle and the alchemists, but to formulate his mechanics he stated that [3, p. 237]:

> ... it seems probable to me that God in the Beginning form'd Matter in solid, massy, hard, impenetrable, movable Particles ...
> For the matter of all things is one and the same, which is transmuted into countless forms by the operations of nature ... and hence we conclude the least particles of all bodies to be also extended, and hard and impenetrable, and movable, and endowed with their proper inertia.

It is clearly implied that all fundamental particles are alike. This proposition has evolved into the modern notion that all elementary particles are zero-dimensional points. Nothing causes more confusion in quantum chemistry than the electronic point particle.

Classical atomism was re-interpreted by John Dalton (1803) as the property of elements, assumed to differ only in atomic weight. His ideas enabled a logical approach to chemical interaction and a captivating stimulus for further research. An immediate aim was to determine the atomic weights and identity of new elements, which rapidly increased in number.

---

[2]Heisenberg and Born arrived at the idea that elementary particles are different quantum states of apeiron—whatever that means.

[3]Modern quantum chemists share this belief as entanglement of hybrid atomic orbitals.

The chemists were so single-minded in their dedication that the brilliant insight of William Prout, that made sense of their blind pursuits, was rejected with such hostility that he had to publish his theory anonymously. Noting the large number of atomic weights on the (H $=$ 1)-scale improbably close to whole numbers Prout hypothesized [4]

> ... that *the atoms of all elements are formed by the condensation of atoms of hydrogen*, this element being the **primary matter** or **protyle**.

It is as if the mule-headed obstinacy that sustained alchemy, against all reason, for two millennia has been driving chemical research ever since. The same obstinacy upheld the phlogiston theory in the face of overwhelming counter-evidence. Even the discoverers of oxygen (Priestly) and hydrogen (Cavendish) refused to consider the experimentally proven alternative theory of combustion. At the present time one marvels at the pathological reluctance to abandon the discredited model of orbital hybridization and linear combination of atomic orbitals as an explanation of chemical interactions.

Prout's hypothesis had the merit of identifying different elements by integer and hence of tying together the concepts of Lucretius (atom), Aristotle (element) and Pythagoras (number) into a single construct.[4] After the discovery of isotopes and the recognition of atomic number Niels Bohr [5] referred to this synthesis as

> ... one of the boldest dreams of natural science ...

Although in denial of its number basis the periodic table of the elements was discovered in the course of the nineteenth century as a function of atomic weight and volume. Together with Balmer's formula the periodic table guided the search for a quantum theory of matter.

Niels Bohr was the first to engage quantum theory in proposing his (1913) planetary atomic model. Unlike Nagaoka he proposed the rotation of a particulate electron in orbit around an atomic nucleus—a single proton in the case of hydrogen. Using classical mechanics and electrostatics he calculated the energy of an orbiting electron as a function of radial distance from the nucleus. In order to match the Rydberg formulation of the Balmer hydrogen spectrum he made the ingenious assumption that the angular momentum of the orbiting electron was restricted to discrete multiples of Planck's constant, in the form of *hbar* ($\hbar = h/2\pi$), which is measured in the same units of *action*. By this definition the integers in the Rydberg formula appear in the Bohr model as *quantum numbers* that quantify the energy, angular momentum and orbital radius of the electron in "stationary quantum states". Any transition of an electron between stationary states of different energy hence results in either emission or absorption of a quantum (photon) of radiation. The *ad hoc* "quantum postulate" of stationary states to represent radiationless motion on "quantized" orbits was necessitated to avoid the problem that an orbitally accelerated charge in orbit must emit radiation and spiral into the nucleus.

---

[4]Only the *void* remained unaccounted for.

The atomic model, without any doubt Bohr's most brilliant contribution to science, is unfortunately physically unacceptable and still responsible for widespread misconceptions about quantum phenomena. Although mathematically more soundly based than Nagaoka's proposal it provided no physical explanation to justify the "quantum" assumption of stationary states, at variance with the laws of electrodynamics. However, the notion of an atom, visualized as a miniature solar system, immediately captured the popular imagination. Despite its glaring defects as a model of three-dimensionally stable electrodynamic systems, it survived as the universal icon of an atom and carries with it the aura of a mysterious quantum reality. It has become fashionable to refer to "the ghost in the atom", to "quantum jumps", and "quantum fluctuations" to convey the impression of some deep understanding, to "explain" unreasonable propositions, and get away with it. Subsequent refinement of the Bohr model contributed very little to dispel the public fascination with quantum magic, and, if anything, added delicious new mysteries with each development.

The physics community accepted Bohr's model provisionally, with the expectation of ultimately finding a suitable interpretation of stationary states. The chemical community remained lukewarm, but for another reason. The ability of circulating charges to link different atoms together during chemical reaction was treated with the scepticism it deserves.

A major advance was due to Arnold Sommerfeld who interpreted the planetary analogy more seriously by proposing Kepler-like elliptic orbits instead of circular Bohr orbits. To achieve this it was necessary to introduce two secondary new quantum numbers, in addition to Bohr's principal quantum number. These numbers implied, not only the quantization of the elliptic eccentricity, but also the spacial orientation of electronic orbits. New quantum rules were required and derived by Sommerfeld from *action integrals*, which involved the product of *conjugate variables* such as momentum and generalized coordinate, and served to generalize Bohr's angular-momentum postulate and provided a *royal road to quantization*.

The practical procedure consists of first describing the electronic motion in terms of continuous classical action variables, followed by quantization, on equating the classical action to an integral multiple of h, the quantum of action. For periodic systems an action-angle integral is evaluated over a complete cycle of 2π, and interpreted as angular momentum, $\hbar = h/2\pi$. The ingenuity of the procedure prompted Sommerfeld to state [6]:

> ... space quantization of the Kepler orbits is without doubt the most surprising result of the quantum theory. The simplicity of the results and their derivation is almost like magic.

Statements like this resonate convincingly with amateur philosophers of science, only too keen to elaborate on the mystical qualities of quantum theory, for general consumption. In reality there is nothing more mysterious about Sommerfeld's result than to Bohr's angular-momentum postulate, of which this is no more than a logical generalization.

However, the introduction of elliptic orbits transformed the unrealistic two-dimensional Bohr model into a more realistic three-dimensional form. The three-dimensional action function, generated by the additional quantum numbers, correctly described the electronic configuration of non-hydrogen atoms, in line with the structure of the periodic table of the elements. The royal road could be used to derive correct quantum rules for molecules and crystals. All molecular spectra, including electronic, vibrational and rotational spectra could be explained in detail and Bragg's diffraction law for periodic crystals could be derived directly. The assignment of electronic spectra in terms of sharp, principal, diffuse and fundamental series, is still recognizable in the s, p, d, f wave-mechanical distinction between electrons, loosely referred to as orbital-angular-momentum vectors.

A schematic drawing of the elliptic orbits that describe the motion of the ten electrons that surround the nucleus of a neon atom shows two circular inner orbits surrounded by a *valence shell* of four circular and four elliptic orbits.

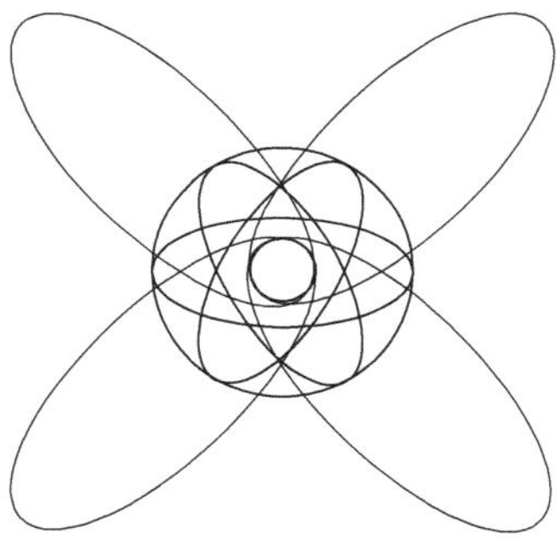

For the carbon atom, with four valence electrons, these were argued to occupy four degenerate elliptic orbits directed towards the corners of a tetrahedron—an arrangement with zero orbital angular momentum.

The predicted tetrahedral structure for CH_4, based on this arrangement was in complete agreement with the Lewis and van't Hoff models and perfectly in line with the theory of covalent bonding. It has been remarked that [7]:

> Theoretical chemistry reached its pinnacle during the Sommerfeld era, before the advent of wave mechanics. The theoretically superior new theory, although it eliminated the paradoxes of zero angular momentum of the hydrogen ground-state, the orbital motion in helium and the nature of the stationary states, it defined the periodic table less well and confused the simple picture of chemical bonding.

Still, to account for elemental periodicity and the Lewis electron-pair model of chemical bonding in terms of Sommerfeld's model it was necessary to empirically add a fourth quantum number, which later on became associated with *electron spin*, without providing a physical explanation of the phenomenon.

At the time when quantum theory found itself in the doldrums and in violation of the laws of electrodynamics, the other new theory of physics was confirmed dramatically during the solar eclipse of 1918.

The theories of relativity were formulated in two instalments. Special relativity developed from several efforts to reconcile the properties of the electromagnetic field, as formulated by Maxwell, with the dynamic theory of relative motion. The sticking point was the electromagnetic requirement of a constant speed of light, irrespective of the relative motion of the light source. A consensual model based on a mathematical formalism, known as the *Lorentz transformation*, eventually came into being as a four-dimensional theory that treats space and time variables as equivalent, in a construct known as *Minkowski space-time*.

A surprising number of scientists and laymen alike, still refuse to accept the implied four-dimensional space-time entanglement. The proposal of alternative new analyses of events, based on trains moving through a station, and communicating with a complicated network of stationary and moving light sources flashing, designed to disprove the theory, has become an international pastime.[5] What these enthusiasts find unpalatable is the demonstration that there is no inexorable universal flow of time; that, like space coordinates, the time measured by different observers depends on their state of motion. Most disturbing is that an event in the future of one observer could be in the past of another, in high-speed relative motion.

All of these efforts are misguided. Assuming the reality of electromagnetism, there is a better chance of improving the theory of relativity by formulating a valid alternative account of electromagnetic phenomena. However, until such time the only option is to accept the space-time model.

One prediction of special relativity, less often in dispute, is the equivalence of mass and energy, as embodied in the best-known equation of physics:

$$E = mc^2 .$$

Not only this equation, but also Einstein's quantum equation that describes a photon, $E = h\nu$, was known to Louis de Broglie, who combined them to read:

$$(p) = mc = h\nu/c = (h/\lambda) .$$

The quantity ν/c defines a wavelength, usually denoted by λ. The product on the left-hand side, in mechanical terms, defines the momentum, p, of a massive object. Once again, Planck's constant, h, is seen to relate a mechanical, or particle, concept to a wave property.

The relationship highlighted by de Broglie is of the same form as Einstein's photon postulate, except that it now deals with a *massive* particle, such as an electron. Whereas Nagaoka defined the electron as a standing wave and Bohr defined it as a particle, de Broglie, following Einstein, opted for the compromise of a dualistic wave-particle composite. He specifically defined an electron as a particle guided by a pilot wave; introducing yet another quantum mystery.

[5]It makes one dizzy to try and follow the convoluted arguments and the proposed sequence of events said to invalidate the theory.

It is remarkable how the same people that refuse to consider a mathematically sound theory, credited with experimentally confirmed predictions of unheard-of physical significance, can enthusiastically support another theory, based on dubious assumptions in conflict with the fundamental laws of physics. This strange schizophrenia is seen perpetuated through the twentieth century. It reflects a tendency closely related to dogmatic belief and obviously inspired by what is considered to be a heretical denial, by the theories of relativity, of the eternal, constant, infinite universe, which is revealed by Newton's religiously inspired celestial mechanics.

Experimental studies of the interaction between radiation and electrons provided conflicting data about the nature of electrons and photons. By studying the scatter of monochromatic hard X-rays on a graphite target A.H. Compton (1923) interpreted his results as confirming the corpuscular nature of electromagnetic radiation. He found the scattered X-rays to consist of two components of different wavelength—a normal component at the same wavelength as the incoming radiation and a second component with the wavelength shifted by an amount that increases with the angle of scatter.

The modified wavelength was ascribed to scatter on electrons that occur practically free in the graphite surface. Using de Broglie's relationship the shift in wavelength translates into a change of photon momentum due to the recoil of the loosely bound electron. Assuming relativistic conservation of momentum and noting the observed angular dependence of the wavelength shift, $\Delta\lambda$, it was shown to depend on the angle of scatter, θ, according to the simple relationship

$$\Delta\lambda = \lambda_C(1 - \cos\theta),$$

where the constant $\lambda_C = h/mc$, known as the *Compton wavelength* of the electron, depends on the relativistic momentum (mc), as predicted by de Broglie's formula in the relativistic limit, ($v \to c$). The quantum energy of radiation with the Compton wavelength is equal to the *rest energy*, m_0c^2, of the electron:

$$h\nu = hc/\lambda_C = m_0c^2\,.$$

Despite Compton's confident characterization of discrete photons the results of his experiments are far from unambiguous. This is obvious from the need to invariably invoke wave parameters in order to account for a supposedly mechanical interaction. The inherent Compton *wavelength* hints at a wave nature, even for the electron. In addition, an equally convincing analysis of Compton scattering, without photons, was described by Schrödinger. An unbiased account of the observation, in contemporary terms, should relate the interaction between an electromagnetic plane wave and a recoiling electron. Instead one finds an obvious paradox confidently dismissed as a *quantum effect*, to resolve the conflict between wave and particle models. Imagine how Zeno would have relished quantum theory.

To complicate matters the phenomenon of electron diffraction was discovered experimentally at about the same time. By then, optical diffraction had been known for a long time and the only credible explanation ever put forward was in terms of a wave model. This observation was therefore in direct conflict with Compton's model of photons and electron particles, by implying a wave nature in both instances.

The quantum confusion was reaching epidemic proportions. How does a wave recoil in collision with another wave? Alternatively, how does a particle diffract? What is a stationary state? How does a pilot wave guide a particle? Where is the boundary between the macroscopic and quantum worlds? What is a photon? In which direction is a photoelectron emitted? How does a point particle spin? How does a wave transfer energy to an orbiting electron? How does radiationless acceleration of an orbiting electric charge work? What is the mechanism of a quantum jump? What principle drives space quantization? How can two particles going in opposite directions on the same orbit avoid collision?

The standard answer at the end of 1924 was simple: All of these queries are meaningless and arise from a misreading of quantum theory!

To put the issue beyond dispute the chase was on to find the details of this magic theory and to the relieve of everybody the problem was cracked within the next two years. But the damage was done. The wild claims about quantum wizardry did not subside, but increased, spread beyond the limits of physics and by now pervades all forms of discourse.

The way in which Sommerfeld's quantization rules could be formulated through action integrals, without reference to orbital motion, served as a guide to the development of a generalized quantum theory without the defects of a system based on a mechanical atomic model.

The classical part of the Sommerfeld action integrals were simply taken over from the Hamilton–Jacobi restructuring of Newtonian mechanics according to Hamilton's principle of least action. An attractive possibility arising from this observation was to modify the HJ theory in its widest sense to generate a non-classical version that incorporates the action as a discrete function.

Hamilton's principle implies that the action, specified by the integral of *kinetic potential* between time limits, has a minimum value for any mechanical process. The integrand, also known as the Lagrangian function, represents the instantaneous difference between the kinetic and potential energies of a mechanical system, formulated in symbols as $L = T - V$. By comparison the Hamiltonian function, defined as $H = T + V$, also represents the time rate of change of *action*; the product of total energy and time, $S = Et$.

By following the progress of a mechanical process that involves a multitude of particles, using the HJ equation, surfaces of constant action are seen to propagate like an infinite wavefront. Conscious of de Broglie's recognition of the wave-like behaviour of electrons, Schrödinger used a general wave equation to solve for a HJ wavefront in terms of the wavelength associated with its quantized mechanical momentum. The resulting Schrödinger wave could be shown to relate to classical particle motion in the same way that wave optics relates to the rays of geometrical optics.

Schrödinger's equation was an immediate sensational success. Not only did it reproduce the Bohr energy levels for hydrogen and the Sommerfeld quantum numbers, without complication, but also the molecular spectra of vibrating and rotating diatomic molecules.

The solutions to Schrödinger's equation are complex wave functions. A complex number is of the form $a + ib$, where i, the square root of -1, is colloquially referred to as *imaginary*. Another source of the myth that nobody can understand quantum theory.

Schrödinger's observation that the wave equation had a form related to the equations of hydrodynamics was developed in detail by Madelung [8] on writing the time-dependence of the solutions, Ψ, as an action function. The linear differential equation may be solved by the separation of space (x) and time (t) variables, into a product function

$$\Psi(x,t) = \psi(x) \cdot e^{i\omega t}$$

in which ω is an angular frequency. In terms of Planck's action constant, another way of formulating the wave function, Ψ, is in the form of an amplitude and a dimensionless complex phase

$$\Psi = Re^{iS/\hbar}.$$

By substituting this solution into Schrödinger's equation Madelung could show how it separates into real and imaginary parts, which closely resemble Euler's coupled differential field equations of hydrodynamics that describe the irrotational flow of a compressible fluid.

To draw the parallel it was necessary to assume that R^2 represented the density of a continuous fluid with stream velocity proportional to the gradient of the action, i.e. $v = \nabla S/m$.

In the case of the hydrogen atom solutions of the time-independent part of the product function, ψ, are also solutions of the general equation, the form used almost exclusively in chemical applications.[6] It correctly defines the stationary states postulated by Bohr and likewise cannot account for transition between different energy states, ascribed by Bohr to quantum jumps. As remarked by Madelung both density and flux for the non-stationary states of the general equation vary with

6

$$\nabla^2\psi + \frac{8\pi^2 m}{h^2}(E - V)\psi = 0.$$

the same periodicity that characterizes the frequency of spectroscopic transitions.[7] This means that radiation is not the result of quantum jumps, but rather happens by slow transition in a non-stationary state.

Characterization of the extranuclear electron density on an atom as equivalent to a periodic continuous fluid, never generally accepted at the time, may turn out to be the most erudite interpretation of wave mechanics after all.

In the theory of atomic spectra periodic systems are of obvious importance and these are handled in HJ theory by the action integrals utilized by Sommerfeld. For systems with many degrees of freedom it leads to the appearance of multiply-periodic functions, conveniently represented in terms of Fourier expansions that involve series of integers, and relate generalized coordinates to frequencies. By fiddling with these integers Heisenberg stumbled onto a special procedure of multiplying and adding generalized coordinates and momenta, to predict the correct observed spectroscopic rules through the Fourier sums.

A surprising feature of Heisenberg's calculus, soon identified as matrix algebra, was its non-commutative nature. This means that a *commutator*, such as $[pq - qp]$, is non-zero. This observation led to Heisenberg's postulate of *quantum uncertainty*, stating that a pair of conjugate variables cannot both have sharply defined values at the same time. One of the reasons that he put forward was that any experimental probe must, of necessity, introduce a disturbance that shows up as an uncertainty. This idea was modified and rephrased several times and in a mysterious way later became part of the equally confusing *quantum measurement problem.*

An unanticipated result, produced by the new quantum models, was the idea of zero-point energy. In the wave-mechanical analysis of particle motion, restricted to limited space, the energy appears quantized as the discrete function of an integral quantum number that also features in describing the wave function associated with the particle. To avoid annihilation of the particle, the wave function, and hence its quantum number, is required to be non-zero, and the lowest energy level must likewise be above zero.

In the case of a harmonic oscillator the definition of zero-point energy more explicitly derives from the general formula that specifies the allowed energy levels as

$$E_n = (n + \tfrac{1}{2})h\nu \quad , \quad E_0 = \tfrac{1}{2}h\nu \, .$$

This phenomenon is without classical analogue, but makes sense as the cause that prevents cooling to absolute 0K. At another level it provides fertile soil for speculation about the nature of the vacuum.

7

$$\nu_{jk} = (E_j - E_k)/h.$$

Heisenberg's result was published a few months before Schrödinger's wave equation, but not surprisingly, the two schemes produced compatible results. Although a satisfactory mathematical solution of the quantum problem was obtained the different physical interpretations of the two schemes are still in dispute.

The strategy that was envisaged to reach consensus on the interpretation of the theoretical results was to arrange for an intensive discussion, involving all prominent contributors and a convenient venue that presented itself was the next of the prestigious Solvay conferences, scheduled for 1927 in Brussels.

The failure to reach consensus as anticipated was due to more than a clash of personalities. The mere fact that the debate still rages almost a hundred years later points at a serious misconception on the part of either, or both, parties; if not a more fundamental philosophical discord.

The quantum debate is not unique in the annals of science. More often than not such dispute can be traced back to a deeper level of discord over issues such as religion, politics and cosmology, or a combination of these. In the present instance the contention is most likely of cosmological nature and related to the philosophical differences between Newton and his continental contemporaries. Before the quantum story continues it is necessary to explore the possible nature of any ideological differences between the opposing camps.

References

1. H. Nagaoka, On the dynamical system illustrating the spectrum lines and the phenomenon of radioactivity. Nature **69**, 392–393 (1904)
2. Lucretius, *The Nature of the Universe* (Penguin Books, Harmondsworth, 1951)
3. A.R. Hall, *The Rise of Modern Science III* (Collins, London, 1963)
4. J.R. Partington, *A Text-Book of Inorganic Chemistry*, 6th edn. (Macmillan, London, 1953)
5. N. Bohr, *Atomic Theory and the Description of Nature* (Cambridge UP, London, 1934)
6. A. Sommerfeld, *Atombau und Spektrallinien*, 4th edn. (Vieweg, Braunschweig, 1924)
7. J.C.A. Boeyens, *Chemistry from First Principles*. Springer.com (2008)
8. E. Madelung, Quantentheorie in hydrodynamischer form. Z. Physik **40**, 322–326 (1926)

Chapter 2
The Cosmological Debate

The newly-born sentient being is confronted with a strange new world that interacts with all of its senses from all directions. For some individuals the effort to interpret the sensations of observation never comes to an end. We all learn from our elders and most people are content with the answers provided by authority and never give it a second thought of their own. They are the salt of the earth—the many-headed that silently flow with the stream.

Another, less populous group, the control freaks, are the self-styled interpreters of the world's collective wisdom, conveniently condensed into easily memorized catch phrases. Most clerics, academics, politicians and scientists belong to this class. They rarely utter an original word and severely criticize or ridicule whoever appears to step out of line. In order to protect this privilege they present a facade of scholarship, proudly supported by extensive research output in the form of millions of indigestible communications.

The secret is to regurgitate the utterances of leading savants, preferably from a bygone era, in slightly modified form. Really "high-class" research would add "new" experimental results to substantiate the ruling paradigm or to offer an apologetic explanation in case of conflict. Research quality is ensured by peer review, the main purpose of which is to prevent the publication of any dissenting point of view.

The driving force behind this mindless activity is financial gain. The importance of a research leader depends on the size of the research group, the number and impact factor of the publications produced and the value of the supporting research grants. This structure remains secure providing favourable peer review is maintained. This means that no dissenting ideas should be allowed to pollute the research output. It is instructive to note how the Pythagorean cult disintegrated when a prominent fellow advocated the recognition of irrational numbers.

Finally there is a third category, traditionally known as philosophers, who never tire of probing for better understanding of the world. Sadly, too many modern-day philosophers have joined the ranks of the scientists in search of authoritarian status.

J.C.A. Boeyens, *The Quantum Gamble*, DOI 10.1007/978-3-319-41621-2_2

We are not concerned with these, but with the much smaller group of scholars, often derogatorily referred to as pseudo scientists, dissidents, crack-pots, quacks or conspiracy theorists, the likes of which, unfortunately, also feature prominently in this category. What all of these have in common is a desire to share a new individual point of view with a disinterested and even hostile establishment. Prime examples from twentieth century science include Alfred Wegener and Immanuel Velikovsky. To disagree with such authors, presenting revolutionary ideas at variance with mainstream models, in a dignified manner is completely acceptable and desirable, but not the hysterical outcry, routinely observed.

The case of Alfred Wegener was taken up by Joseph Sant, who wrote [1]:

> Alfred Wegener was the scientist who championed the Continental Drift Theory in the early twentieth century. Simply put, his hypothesis proposed that the continents had once been joined, and over time had drifted apart. The jigsaw fit that the continents make with each other can be seen by looking at any world map.
> Since his ideas challenged scientists in geology, geophysics, zoogeography and paleontology, it demonstrates the reactions of different communities of scientists. The reaction from the different disciplines was so negative that serious discussion of the concept stopped. One noted scientist, the geologist Barry Willis, seemed to be speaking for the rest when he said:
>
> > "... further discussion of it merely incumbers the literature and befogs the mind of fellow students."
>
> Barry Willis's and the other scientists wishes were fulfilled. Discussion did stop in the larger scientific community and students' minds were not befogged. The world had to wait until the 1960s for a wide discussion of the Continental Drift Theory to be restarted.
> Why did Alfred Wegener's work produce such a reaction? He was diplomatic in presenting his theory. Although he believed some of his arguments were compelling, he knew he would need more support to convince others. His immediate goal was to have the concept openly discussed. Wegener did not even present Continental Drift as a proven theory. These modest goals did not spare him. His work crossed disciplines. The authorities in the various disciplines attacked him as an amateur that did not fully grasp their own subject. More importantly however, was that even the possibility of Continental Drift was a huge threat to the established authorities in each of the disciplines.
> One can't underestimate the effect of a radical new viewpoint on those established in a discipline. The authorities in these fields are authorities because of their knowledge of the current view of their discipline. A radical new view on their discipline could be a threat to their own authority. One of Alfred Wegener's critics, the geologist R. Thomas Chamberlain, could not have summarized this threat any better:
>
> > "If we are to believe in Wegener's hypothesis we must forget everything which has been learned in the past 70 years and start all over again."
>
> He was right.

When I first heard about continental drift as a primary-school kid, not only the matching of outlines but also of geological features, such as mountain ranges of the same type, between South Africa and South America, convinced me of the validity of the argument that baffled the experts.

However, when the same concept was rediscovered thirty years later as the theory of plate tectonics it was universally accepted without argument. The only difference is that the new theory explicitly formulated the drift as a slow process continued over millions of years, while Wegener's proposal was suspect as possibly implying some catastrophic event. A prospect too frightening to contemplate for the establishment and in heretical contradiction of Newton's authority.

Isaac Newton is universally remembered as the author of *Principia*, of which it has been said [2]:

> No other work in the whole history of science equals the *Principia* either in originality and power of thought, or in the majesty of its achievement. No other so transformed the structure of science, for the *Principia* had no precursor in its revelation of the depth of exact comprehension that was accessible through mathematical physics. No other approached its authority in vindicating the mechanistic view of nature, which has been so far extended and emulated in all other parts of science. There could be only one moment at which experiment and observation, the mechanical philosophy, and advanced mathematical methods could be brought together to yield a system of thought at once tightly consistent in itself and verifiable by every available empirical test. Order could be brought to celestial physics only once, and it was Newton who brought order.

Who could be rash enough to challenge such authority? Nobody would, without realizing that [2]:

> Newton ... kept his views ... [on religion] ... to himself, though as was his habit, he poured out thousands of words and notes and draft essays which he never intended for any eye but his own

Apart from religion these unpublished works also deal with alchemy, politics, and ancient history. According to Frank Manuel [3] the main purpose of this material was to discredit all the historical evidence presented for changes in the solar system. Cohen [4] agrees with this assessment, stating that

> "Newton had one real secret, and concerning it he did his best to keep the world in ignorance."

The secret is concluded to be an intention to uphold the theology and cosmology of Maimonides,[1] in agreeing with Aristotle that

> "... the cosmos, once created is permanent and indestructible."

Most biographers portray Newton as a dedicated scientist, often in conflict with contemporaries, such as Robert Hooke, over the interpretation and principles of Science. More detailed recent studies of Newton's unpublished memoirs and the writings of colleagues, such as William Whiston, expose a completely different, more sinister, side of Newton's philosophies, beliefs and cosmology.

As related by Stecchini [5], Newton's interest in cosmology was aroused by the work of William Whiston, whom he recommended as his successor on his retirement in 1703. In Whiston's acclaimed *New Theory of the Earth*, he contended *inter alia* that [5]:

> ... the cataclysm described in the Old Testament as universal Deluge was caused by the impact of a comet ...

[1] Who branded Christianity as idolatrous.

This conjecture was supported by Edmund Halley and critically reviewed in a book by John Keill [6]. In dedicating the book to the Master of Balliol College it is stated that:

> ... the design of it is no other than to shew, that true Philosophy doth not contradict the Scriptures ...

After stating:

> ... I cannot but acknowledge that Mr. Whiston ... has made great discoveries ... In his theory there are some coincidents which make it indeed probable, that a comet at the time of the Deluge passed by the earth.

He continues:

> ... notwithstanding this, I believe ... that a Comet could never have produced those various effects that Mr Whiston has attributed to it; and it will also further appear that the Deluge was the immediate work of the Divine power ...
> ... I think Mr. Whiston has not exactly observed his first *Postulatum, viz,* that the obvious and literal sense of Scripture is the true and real one ...
> I know indeed that there are some, who are ... for explaining ... most of the ... extraordinary events recorded in the holy Scripture, by natural principles: But I dare suppose Mr. Whiston would not willingly be put into a Catalogue with such Authors.

By 1713, when the second edition of *Principia* was published, Newton's feelings towards Whiston had changed to the point where he threatened to resign from the presidency of the Royal Society should Whiston be admitted as a member.

Most commentators agree that Newton's reaction against Whiston resulted from long-held religious convictions, as evidenced by theological preoccupations that dominate the second edition of *Principia* [7]. He argued that the world was created by a single act and remained stable and unchanged since creation. To account for gravitational distortion of planetary orbits he claimed that

> ... God in his providence must intervene from time to time to reset the clockwork of the heavens to its original state.

Despite contemptuous rejection of such views by Leibniz, Laplace and others, a flood of popularizations elevated Newton's view into the basic doctrine of the eighteenth century. Newton confessed in his Scholium:

> This most beautiful system of sun, planets and comets, could only proceed from the counsel and dominion of an intelligent and powerful Being.

As commented by Herbert Butterfield [8]:

> The ideal of a clockwork universe was the great contribution of the seventeenth century to the eighteenth-century age of reason.

John Maynard Keynes, one of the first to examine Newton's secret papers, commented on '*Newton, the Man*', in an essay prepared for Newton's tercentenary [9], as follows:

> In the eighteenth century and since, Newton came to be thought of as the first and greatest of the modern age of scientists, a rationalist, one who taught us to think on the lines of cold and untinctured reason.

> I do not see him in this light.
> Very early in life Newton abandoned orthodox belief in the Trinity. ... He was rather a Judaic monotheist of the school of Maimonides.
> But this was a dreadful secret which Newton was at desperate pains to conceal all his life.
> It is a blot on Newton's record that he did not murmur a word when Whiston, his successor in the Lucasian chair, was thrown out of his professorship and out of the University for publicly avowing opinions which Newton had secretly held for upwards of fifty years past.

Newton's hostility most probably contributed to Whiston's dismissal from his teaching position because of heresy and subsequent trial before the bishops of the Church of England.

It is becoming increasingly evident from his unpublished manuscripts that science *per se* was not Newton's main interest. He appears as a biblical fundamentalist whose brilliant scientific endeavours were intended to reconcile science and astronomy with his religion, in line with the medieval synthesis, by Augustine and other saints, of biblical religion with Aristotelian cosmology. Despite enlightened opposition Newton's bigotry carried the day on the back of his colossal reputation and survived into the twentieth century.

Growing present-day dissatisfaction with the peer-review system is ascribed to the stranglehold of an elitist establishment on communication and financial resources pertaining to research. Who are this elite? In the Western World they are the heirs of Newton's bigotry and their influence is sufficient to control (or distort) the practice of science worldwide. Their allegiance to religious doctrine, like that of Newton, never openly articulated, is diligently practiced as unruffled support in upholding the lofty ideals of scientific integrity.

Those adversely affected lament the decline of science into a dogmatic pseudo religion. They are mistaken. Science, practiced in the Newtonian spirit, is a religion. Modern-day Whistons are tolerated providing they respect the doctrines of quantum mechanics and chemistry, particle physics, big-bang cosmology, string theory, Darwinian evolution, archeology and politics. A well-documented example of the twentieth-century holy Inquisition of Science in action, is the so-called Velikovsky affair.

Immanuel Velikovsky had the temerity to revive the catastrophe theories of Whiston, Donnelly [10] and others at a time when a spirit of intolerance swept the USA. He re-examined the conjecture that anciently reported upheavals such as deluges and other miraculous events could have been caused by a close encounter between a massive comet and planet earth. He scoured ancient writings, traditions and mythical sagas as evidence for reconstructing a speculative scenario, based on interplanetary interactions that occurred in historical times, in order to account for the purported events. To account for the frequent mention of *thunderbolts* he made the bold suggestion that electromagnetic interactions are of major importance, in addition to gravity, to account for cosmic phenomena.

To the annoyance of the scientific establishment his book, *Worlds in Collision*, 1950, was an immediate best-seller. The interdisciplinary nature of the background research presented an ideal platform from which to discredit the work. Critics and reviewers from all disciplines had a field day, but the charge was led by the Harvard astronomers Harlow Shapley and Carl Sagan.

The first effort to suppress the book was partially successful in so far as persuading the publishers, by a threatened academic boycott, to transfer the rights to another house. Only then did the serious scientific onslaught commence. The most telling point was to attack the credentials and integrity of the author. However, the real threat that had to be destroyed was the notion of catastrophism and the role of electromagnetism in cosmic events, the former openly contradicting Newton's doctrine, which can be traced back through Aristotle to Plato's idea of the eternal regularity of the heavenly bodies [11], Book X:

> In the first place, the earth and the sun, and the stars and the universe, and the fair order of the seasons, and the division of them into years and months, furnish proof of [the] existence of the Gods.
> The ruler of the universe has ordered all things with a view to the excellence and preservation of the whole...

Those not in agreement should

> ... be placed by the judge in the House of Reformation, and ordered to suffer imprisonment during a period of not less than five years. And in the meantime let them ... converse with a view to the improvement of their soul's health. [...] if condemned a second time, let him be punished with death.

Details about the hysteria among academics following the publication of Velikovsky's works have been analyzed many times, e.g. [12, 13] and will not be repeated here. It is more important to note how discoveries from the space-exploration programmes continue to vindicate some views for which Velikovsky was ostracized. Most spectacular among these were the discovery of the Van Allen radiation belts, the high surface temperature of Venus, the high-potential negative charge on the sun, radio noise eminating from Jupiter and the extent of interplanetary magnetic fields. Even the discrepancy between ancient chronologies, identified by Velikovsky, is now supported by more recent scholarship.

Newton's cosmology is generally upheld in the West although it remains silent about the way in which the cosmos first came into existence. No effort is made to interpret creation myths and until very recently conjectures, such as Edgar Allan Poe's *Eureka*, had been treated as science fiction [14]. All of that changed abruptly at the end of the second world war. Not only did science fiction become science, but it soon blossomed into "big science", with all the necessary trimmings of absolute conviction and intolerance of dissident views.

Instead of a cosmic egg or a mighty verbal command it was a singularity that gave birth to the universe, some 14 billion years ago. The popular version contends that this Big-Bang theory is based on the theory of general relativity and supported by all astronomical evidence. The theory of general relativity is too poorly understood by both protagonists and dissidents alike, for them to seriously enter into an argument, although Big Bang theory has demonstrably nothing in common with the theory of general relativity [15, 16]. The only vulnerability of the concept is from astronomical observation against which it is ruthlessly defended. An iconic victim of the persecution for such an offense is the famous astronomer Halton Arp, who on occasion had galaxies named after him.

The fundamental assumption of BB cosmology is that the observed spectroscopic redshift of galactic light is a measure of distance. Arp's indiscretion was to observe physical links between galactic objects with massively different redshifts. This invalidates the concept of an expanding universe that underpins the BB.

In retaliation Arp lost his teaching position, was denied any further telescope time and had to move to Germany's Max Planck Institute in order to carry on with his work. Many journals are reluctant to publish Arp's papers, under the influence of cosmologists claiming that all his observations had been explained in full. Despite accumulating evidence in support of Arp's doubts, he will probably have to wait, like Copernicus, for posthumous recognition.

Response to the publication of Schrödinger's papers on wave mechanics was immediate and lethal. A revanchist cabal laid into the author and his work with equal ferocity. The most mischievous accusation blamed him for a secret desire to undo the quantum revolution and return to the classical system. Not only was the man reviled as a womanizer, but also for being in love with waves.

Although never mentioned explicitly, it was the Newtonian particle model that had to be protected at all costs. The concept of quantum probability was concocted for this very purpose and to simulate wave motion in the particle model. To get around the phenomenon of electron diffraction they expressed lukewarm support of de Broglie's proposal, but eventually championed the ludicrous wave-particle duality. In reality, the more precise cabalistic view was of a **particle** with wave properties as needed.

One looks in vain for an understanding of dogmatic science. Why does an ostrich bury its head in the sand to avoid an unpleasant reality? Why could the bishops distinguish no features on the lunar surface when looking through Galileo's telescope? What prevented trained geologists to appreciate the obviously matching continental shorelines pointed out by Wegener? Why ostracize scholars like Velikovsky and Arp for offering alternative views?

The chickens, regrettably, very seldom come home in time to roost. Dissident scientists are, too often only vindicated by a next generation. The evidence is accumulating that Schrödinger's time has come. The time has arrived to reopen the quantum debate. For a start it will be necessary to have an unbiased remorseless examination of all evidence before a convincing case can be made out for adopting a new quantum theory. The probability gamble has failed and should be put to the sword.

Starting from Schrödinger and Madelung's hydrodynamic model, the road beckons through GTR and the topology of space-time, back to the higher-dimensional hydrodynamic models of nonlinearity, hypercomplex algebra and number theory. We sense a window with a view on yet another exciting new step forward in science.

References

1. J. Sant (2014), *Wegener and Continental Drift Theory*. http://www.scientus.org/Wegener-Continental-Drift.html
2. A.R. Hall, *From Galileo to Newton*: The Rise Of Modern Science·III (Collins, London, 1963)
3. F. Manuel, *Isaac Newton, Historian* (Cambridge, 1963)
4. I.B. Cohen, Franklin and Newton. Am. Phil. Soc. **43**, (1956)
5. L.C. Stecchini, *The Inconstant Heavens*, in [12]
6. J. Keill, *Some remarks on Mr. Whiston's Theory of the Earth* (Clements, London, 1734)
7. F. Cajori, *An Historical and Explanatory Appendix* to *Edited version of Principia* (Berkeley, 1946)
8. H. Butterfield, *The Origins of Modern Science* (NY, 1960)
9. http://www-history.mcs.st-andrews.ac.uk/Extras/Keynes-Newton.html
10. I. Donnelly, *Ragnarok: The Age of Fire and Gravel*, (1883); (Nabu Press, 2010)
11. Plato, *Laws*, The Project Gutenberg EBook #1750, Transl. by B. Jowett (2008)
12. A. de Gracia (ed.), *The Velikovsky Affair: Scientism vs Science* (Metron Publications, Princeton, NJ, 1978)
13. C. Ginenthal, *Carl Sagan and Immanuel Velikovsky* (New Falcon Publications, 1995)
14. H. Beaver (ed.) *The Science Fiction of Edgar Allan Poe* (Penguin, Harmondsworth, 1976)
15. J.C.A. Boeyens, *Cosmology and Science*, in [16]
16. A. Travena and B. Soren, *Recent Advances in Cosmology* (Nova, New York, 2013)

Chapter 3
What Happened . . .

It has been common knowledge for decades that the authentic interpretation of quantum theory, after being threshed out in detail at the 1927 Solvay Conference, was finalized and documented in Copenhagen in a form supported by all competent physicists. The ultimate generalized version which developed in the USA after the second world war is widely considered as the most successful scientific theory ever formulated, but summarized by Dirac in one word: "ugly". Despite the euphoria of quantum physicists all efforts to export the theory as a basis for theoretical chemistry and cosmology have failed dismally. It could well be that in the race to produce the perfect theory something was overlooked. Could it be in Brussels, in Copenhagen or the USA? The only way to tell is by an unbiased scrutiny of all steps in the process.

In Brussels?

Of particular importance, especially for pedagogues, were the few written accounts of the Solvay Proceedings, dutifully transcribed and elaborated in countless textbooks. The official French version of the proceedings was forgotten and gathered dust in some archive, until it was excavated, translated into English and published with scholarly comments [1].

What comes as a complete surprise is the way in which the official record of events flatly contradicts all popular accounts. In review of this work, L.L.Williams quotes as follows from his own, previously published, distorted account:

> At the 1927 Solvay Congress. . .de Broglie presented a pilot wave interpretation of the wave function. De Broglie suggested that the wave function was a real physical field that affected the motion of material objects. He proposed that it was a discovery of a kind with the electric and gravitational fields. Indeed, the Schroedinger equation can be recast to reflect this view. [true so far] De Broglie's interpretation withered under the direct objections raised by Pauli which he was unable to convincingly repudiate. [not true] It expired under the influence of Schroedinger and Einstein. . .[not true].

J.C.A. Boeyens, *The Quantum Gamble*, DOI 10.1007/978-3-319-41621-2_3

After giving several further examples of misrepresentation of the events in Brussels the review closes with the lament:

> The mystery is why all this was lost to the mainstream history of science until now.

Fact is that the proceedings, under the title: *Électrons et Photons*, was only published a year later, while biased accounts started circulating immediately.

One of the most outrageous accounts is documented as follows [1]:
"One frequently cited piece of contemporary evidence is a description of the conference written by Ehrenfest a few days later, in a letter to his students and associates in Leiden. An extract reads:

> Bohr towering completely over everybody ... step by step defeating everybody ... It was delightful for me to be present during the conversations between Bohr and Einstein ... Einstein all the time with new examples. In a certain sense a sort of Perpetuum Mobile of the second kind to break the UNCERTAINTY RELATION. Bohr ... constantly searching for the tools to crush one example after the other. Einstein ... jumping out fresh every morning. ... I am almost without reservation pro Bohr and contra Einstein."

Not surprisingly [1]:

> The fifth Solvay conference is usually remembered for the clash that took place between Bohr and Einstein, supposedly concerning in particular the possibility of breaking the uncertainty relations. It might be assumed that this clash took the form of an official debate that was the centrepiece of the conference. However, no record of any such debate appears in the published proceedings, where both Bohr and Einstein are in fact relatively quiet.
> The available evidence shows that in 1927 the famous exchanges between Bohr and Einstein actually consisted of informal discussions, which took place semi-privately (mainly over breakfast and dinner), and which were overheard by just a few participants, in particular Heisenberg and Ehrenfest. The historical sources for this consist, in fact, entirely of accounts by Bohr, Heisenberg and Ehrenfest.

Probably more than anything else, the widely respected books of Jammer [2, 3] perpetuated a false impression of the Solvay events, with statements such as:

> Bohr's masterly report of his discussions with Einstein on this issue, though written more than 20 years after they had taken place, is undoubtedly a reliable source for the history of the episode.

There goes the theory that Bohr triumphed over Einstein.

Jammer further declares that:

> [de Broglie's theory] was hardly discussed at all... the only serious reaction came from Pauli...
> It was immediately clear that nobody accepted his ideas ... In fact, with the exception of some remarks by Pauli ... de Broglie's causal interpretation was not even further discussed at the meeting. Only Einstein once referred to it *en passant*.

To put the record straight the conference was centred around five invited contributions, each followed by extensive discussion, and concluded with a free-ranging final discussion. Three of the presentations, those by de Broglie, by Born and Heisenberg, and by Schrödinger, are directly concerned with the interpretation of quantum events. Bohr did not make a contribution, as often claimed, and although he offered a written contribution, it was not accepted as part of the proceedings. It has been emphasized by way of introduction [1] that:

> The interpretation of quantum theory seems as highly controversial today as it was in 1927. There has also been criticism—on the part of historians as well as physicists—of the tactics used by Bohr and others to propagate their views in the late 1920s, and a realisation that alternative ideas may have been dismissed or unfairly disparaged. For many physicists, a sense of unease lingers over the whole subject. Might it be that things are not as clear-cut as Bohr and Heisenberg would have us believe? Might it be that opponents had something important to say after all?

Careful analysis of the proceedings [1] shows that discussion centred around the following contentious issues:

1. Is the electron a point particle with continuous trajectory (de Broglie), or a wave packet (Schrödinger), or neither (Born and Heisenberg)?
2. Do quantum outcomes occur when nature makes a choice (Dirac), or when an observer decides to record them (Heisenberg)?
3. Is the non-locality of quantum theory compatible with relativity (Einstein)?
4. Is indeterminism a fundamental limitation, or merely the outcome of coarse-graining over something deeper and deterministic (Lorentz)?

The debate around these points in open discussion was muted, but fierce. There is little evidence of approaching consensus on any of the issues, as reflected in a selection of various responses and comments scattered throughout the text [1].

1. **Particle and/or Wave?**

   From the outset the possibility of reaching consensus on the particle–wave issue was close to zero. The frame of mind in which the protagonists arrived in Brussels is evident from some indirect exchanges.

   > In a much-quoted footnote, Schrödinger says he had known of Heisenberg's theory but was 'scared away, not to say repelled', by the complicated algebraic methods and the lack of *Anschaulichkeit*.

   At about the same time Heisenberg wrote to Pauli:

   > The more I reflect on the physical part of Schrödinger's theory, the more disgusting I find it.... What Schrödinger writes about *Anschaulichkeit* makes scarcely any sense, in other words I think it is bullshit [Mist].

   - **De Broglie's Position**

     De Broglie's contention according to his first communication in 1923 was

     > ...that an 'internal periodic phenomenon' should be associated with *any* massive particle (including light quanta).
     > He proposes that orbits are stable only if the fictitious wave remains in phase with the internal oscillation of the electron. From this condition, de Broglie derives the Bohr–Sommerfeld quantisation condition.

     He defended the same point of view in his Nobel lecture:

     > ... the determination of the stable motions of an electron in the atom involves whole numbers, and so far the only phenomenon in which whole numbers were involved

> in physics were those of interference and eigenvibrations. That suggested the idea to me that electrons themselves could not be represented as simple corpuscles either, but that a periodicity had to be assigned to them too.

To put this into perspective

> [i]t is sometimes claimed that, for the case of electrons, ideas similar to de Broglie's were put forward by Madelung in 1926. What Madelung proposed, however, was to regard an electron with mass m and wave function ψ not as a point-like particle within a wave, but as a continuous fluid spread over space with mass density $m|\psi|^2$. In this 'hydrodynamical' interpretation, mathematically the fluid velocity coincides with de Broglie's velocity field; but physically, Madelung's theory seems more akin to Schrödinger's theory than to de Broglie.

It is important to note that

> [a] common historical misconception concerns the reception of de Broglie's theory at the Solvay conference. It is usually said that de Broglie's ideas attracted hardly any attention. It is difficult to understand how such an impression originated, for even a cursory perusal of the proceedings reveals that de Broglie's theory was extensively discussed, both after de Broglie's lecture and during the general discussion.

The confrontation with Pauli originated in a letter where

> ... Pauli states that he is suspicious of de Broglie's trajectories and cannot see how the theory could account for the discrete energy exchange seen in individual inelastic collisions between electrons and atoms. ... This is essentially the objection that Pauli raises less than three months later in Brussels.
> In the general discussion Pauli begins by stating his belief that de Broglie's theory works for elastic collisions.
> Pauli goes on to claim that de Broglie's theory will not work for inelastic collisions, in particular for the example of scattering by a rotator. ... Pauli's objection is framed in terms of Fermi's analogy, and therein lies the confusion.

About de Broglie's response:

> While [he] was misled by Fermi's analogy, even so his remarks contain the key point, later developed in detail by Bohm—that finiteness of the initial packet will ensure a separation into non-overlapping final packets in configuration space.
> Finally we point out that—leaving aside the misleading nature of Fermi's optical analogy—de Broglie's audience may well have understood his description of wave packets separating in configuration space, for Born had already described something very similar for the Wilson cloud chamber, earlier in the general discussion.
> ... de Broglie's reply to Pauli's criticism contained the essential ideas needed for a proper rebuttal. ... the claim that de Broglie abandoned his theory because of Pauli's criticism is also not true.

- **Heisenberg's Interpretation**
  At the conference Born and Heisenberg criticized both de Broglie's pilot wave and Schrödinger's wave mechanics, without explicitly stating their own interpretation, which is not clearly distinguished from their probability model.
  Schrödinger's concepts are dismissed with contempt:

> [Schrödinger']s view...that these waves exhaust the essence of matter and that particles are nothing but wave packets, not only stands in contradiction with

> the principles of Bohr's empirically very well-founded theory, but also leads to impossible conclusions; here therefore it shall be left to one side.

Using the demonstrated equivalence of matrix and wave mechanics they claimed that Schrödinger's equation was "a special case of the operator theory", derived from their commutation rules. They illustrate their basic philosophy in terms of quantum jumps:

> If one asks the question *when* a quantum jump occurs, the theory provides no answer. At first it seems as if there were a gap here which might be filled with further probing. But soon it became apparent that this is not so, rather, it is a failure of principle, which is deeply anchored in the nature of the possibility of physical knowledge.

Their presentation is concluded by the statement which is still quoted almost verbatim by the most devoted partisans:

> By way of summary, we wish to emphasise that while we consider the last-mentioned enquiries, which relate to a quantum mechanical treatment of the electromagnetic field, as not yet completed [unabgeschlossen], we consider *quantum mechanics* to be a closed theory [gesclossene Theorie], whose fundamental physical and mathematical assumptions are no longer susceptible of any modification. Assumptions about the physical meaning of quantum mechanical quantities that contradict Jordan's or equivalent postulates will in our opinion also contradict experience. (Such contradictions can arise for example if the square of the modulus of the eigenfunction is interpreted as charge density).
> ... the dualism between corpuscles and waves, which in quantum mechanics appears as part of a contradiction-free, closed theory [abgeschlossene Theorie], holds in quite a similar way for radiation. The relation between light quanta and electromagnetic waves must be just as statistical as that between de Broglie waves and electrons.

Weeks before the conference Pauli wrote to Bohr:

> In the last number of the Journal de Physique, a paper by de Broglie has appeared ... de Broglie attempts here to reconcile the full determinism of physical processes with the dualism between waves and corpuscles ... even if this paper by de Broglie is off the mark (and I hope that actually), still it is very rich in ideas and very sharp, and on much a higher level than the childish papers by Schrödinger, who even today still thinks he may ... abolish material points.

About de Broglie's proposals, during the conference, is stated:

> The dual nature of light—waves, light quanta—corresponds to the analogous dual nature of material particles; these also behave in a certain respect like waves.

Their own position is described as follows:

> ... we attribute a dual nature to matter also; its description requires both corpuscles (discontinuities) and waves (continuous processes). From the viewpoint of the statistical approach to quantum mechanics it is now clear why these can be reconciled: the waves are probability waves. Indeed, it is not the probabilities themselves, rather certain 'probability amplitudes' that propagate continuously and obey differential or integral equations, as in classical continuum physics; but additionally there are

> discontinuities, corpuscles whose frequency is governed by the square of their amplitudes.[1]

They explain:

> The corpuscular structure of light thus appears here as quantisation of light waves, similarly to how vice versa the wave nature of matter manifests itself in the 'quantisation' of the corpuscular motion.
>
> The essence of electrons thus plays the same role in the formal elaboration of the theory as that of light quanta; both are discontinuities no different in kind from the stationary states of a quantised system. However, if the material particles stand in interaction with each other, the development of this might run into difficulties of a deep nature.

It is emphasized [1] that:

> ... Born and Heisenberg's report contains some remarkable comments about the nature of interference. ... [Early in their presentation] they explicitly assert that a system always occupies a definite energy state at any one time, while in the later treatment of arbitrary observables nothing is said about whether a system always possesses definite values or not.
> ... Born and Heisenberg resolved the difficulty by asserting that unmeasured population probabilities are somehow not applicable or meaningful. This does not seem consistent with the view they expressed earlier, that atoms always have definite energy states even when these are not measured. How could an ensemble of atoms have definite energy states, without the energy distribution be meaningful?

- **Schrödinger Waves**
  Schrödinger's model is well defined and concisely summarized as follows [1]:

> Schrödinger's work on wave mechanics in 1926 appears to have been driven by the idea that one could give a purely wave-theoretical description of matter. Key elements in this picture were the idea of particles as wave packets and the possible implications for the problem of radiation. This pure wave theory, in contrast to de Broglie's work, did away with the idea of particle trajectories altogether. The main conflict, however, was between Schrödinger and the proponents of quantum mechanics (in particular Heisenberg), both in its form at the time of Schrödinger's papers and in its further developments...
> The idea of particles as wave packets is crucial to the development of Schrödinger's ideas and appears to provide one of the main motivations, at least initially, behind the idea of a description of matter purely in terms of waves.
> Schrödinger's earliest speculation about wave packets (for both material particles and light quanta) is found in his paper on Einstein's gas statistics [4].
> In [a] letter to Planck he admits that there will always exist spread-out states, because of linearity, but still hopes it will be possible to construct packets that stay together for hydrogen orbits of high quantum number.
> ...[in his] paper on micro- and macromechanics...he showed that for the harmonic oscillator, wave packets do stay together.
> The problem of the relation between micro- and macro physics is connected of course to the linearity of the wave equation, which appears to lead directly to highly

[1] The nature of corpuscular frequency is not elucidated any further.

non-classical states (witness Schrödiger's famous 'cat' example). One might ask whether Schrödinger himself considered the idea of a non-linear wave equation. In print, Schrödinger mentioned the possibility of a non-linear term in order to include radiation reaction in his fourth paper on quantisation.

In a letter to Schrödinger

> Lorentz ... pointed out that while beats would arise if the time-dependent wave equation were linear, they would still not produce radiation by any known mechanism. Combination tones would arise if the wave equation were non-linear. ... once the charge density of a particle is associated with a quadratic function of ψ, such as $|\psi|^2$, 'difference tones' in the oscillating charge density arise regardless of whether the wave equation is linear or non-linear.

An important point with respect to any possible model of matter waves was made by de Broglie:

> ... de Broglie criticise[d] Schrödinger's interpretation, according to which a particle identified with an extended, non-singular wave packet, having no precise position or trajectory. De Broglie objects that, without well-defined particle positions, the coordinates $\mathbf{x_1}, \ldots, \mathbf{x_N}$ used to construct the configuration space is 'purely abstract', and that a wave propagating in this space cannot be a physical wave: instead, the physical picture of the system must involve N waves propagating in 3-space

2. **Quantum Measurement**

The question of associating quantum outcomes with a choice made by either nature or an observer came up during the general discussion. Dirac made the point that

> According to quantum mechanics the state of the world at any time is describable by a wave function ψ, which normally varies according to a causal law, so that its initial value determines its value at any later time. It may however happen that at a certain time t_1, ψ can be expanded [as a linear combination of ψ_n's] where the ψ_n's are wave functions of such a nature that they cannot interfere with one another at any time subsequent to t_1. If such is the case, then the world at times later than t_1 will be described not by ψ but by one of the ψ_n's. The particular ψ_n that it shall be must be regarded as chosen by nature. One may say that nature chooses which ψ_n it is to be ...

In response Heisenberg replied:

> I should rather say ... that the *observer himself* makes the choice, because it is only at the moment when the observation is made that the 'choice' has become a physical reality and that the phase relationship in the waves ... is destroyed.

3. **Non-locality**

The few remarks in the Solvay Proceedings that can be interpreted as related to Einstein's putative debate with Bohr have been shown conclusively [1] not to be on *uncertainty* at all, but about non-locality. It is amusing to note that Bohr did not understand what Einstein was talking about, as became clear again later on after publication of the famous EPR paper. This is confirmed by the excerpts below. The concept of *entanglement* used in the discussion differs from its use in relativistic context where it refers to inseparability of space and time coordinates.

> ... as shown in detail by Howard [5], the real nature of Einstein's objections was in fact misunderstood by Bohr, Heisenberg and Ehrenfest—for Einstein's main target was not the uncertainty relations, but what he saw as the non-separability of quantum theory.
> Howard argues ... that, in spring 1927, Einstein must have realised that Schrödinger's wave mechanics in configuration space violated separability, because a general solution to the Schrödinger equation for a composite system could *not* be written as a product over the components. [...] Einstein objected to entanglement, and concluded that the wave function in configuration space could not represent anything physical.
> The primary aim of the famous thought experiments that Einstein discussed with Bohr ... was not to defeat the uncertainty relations but to highlight the (for him disturbing) feature of quantum theory—that spatially separated systems cannot be treated independently.
> ... Einstein's criticism in the general discussion concerned locality and completeness (just like the later EPR argument), not the uncertainty relations. Bohr, in his reply, states that he does not 'understand what precisely is the point' Einstein is making. It seems rather clear that, indeed, in 1927 Bohr did not understand Einstein's point, and it is remarkable that what is most often recalled about the fifth Solvay conference was in fact largely a misunderstanding.
> The key point here is that, if there is no action at a distance, and if the extended field ψ is indeed a complete description of the physical situation, then if the electron is detected at a point P on the film, it could happen that the electron is also detected at another point Q, or indeed at any point where $|\psi|^2$ is non-zero. Upon detection at P, it appears that a 'mechanism of action at a distance' prevents detection elsewhere.
> Einstein's argument is so concise that its point is easily missed, and one might well dismiss it as arising from an elementary confusion about the nature of probability. (Indeed, Bohr comments that he does not 'understand what precisely is the point' Einstein is making).

4. **Indeterminacy**

The notion of indeterminacy was introduced by Born and Heisenberg. It is pointed out that [1]:

> ... Born and Heisenberg's contribution at the Solvay conference do not seem sufficiently clear or complete to warrant definite conclusions as to what they believed concerning the precise relationship between probabilities and the wave function ...

Here is an example of how they are skirting around the issues:

> ...application of the old (classical) words and concepts is allowed, such as 'position, velocity, momentum, energy of a particle (electron)'. It now turns out that all these quantities can be *individually* exactly measured and defined, as in the classical theory, but that for simultaneous measurements of canonically conjugate quantities (more generally: quantities whose operators do not commute) one cannot get below a characteristic limit of indeterminacy.

Not a word about probability. In fact, internal tension between Born's *probabilities*, Heisenberg's *uncertainties* and Bohr's *complementarity* always lurks just below the surface:

> ... difference in opinion was reflected in the sometimes tense discussions between Heisenberg and Bohr at the time, particularly on the topic of Heisenberg's treatment of the γ-ray microscope in the uncertainty paper and the corresponding addendum in proof.

The question of indeterminism was reopened by the conference chairman, H.A. Lorentz, during the general discussion:

> MR LORENTZ—[...] I imagine that, in the new theory, one still has electrons. It is of course possible that in the new theory, once it is well-developed, one will have to suppose that the electrons undergo transformations. I happily concede that the electron may dissolve into a cloud. But then I would try to discover on which occasion this transformation occurs. If one wished to forbid me such an enquiry by invoking a principle, that would trouble me very much. It seems to me that one may always hope one will do later that which we cannot yet do at the moment. Even if one abandons the old ideas, one may always preserve the old terminology. I should like to preserve this ideal of the past, to describe everything that happens in the world with distinct images. I am ready to accept other theories, on condition that one is able to re-express them in terms of clear and distinct images.
> ... to explain the phenomena, it suffices to assume that the expression $\psi\psi^*$ gives the probability that the electrons and photons exist in a given volume [...] But the examples given by Mr Heisenberg teach me that I will have thus attained everything that experiment allows me to attain. However, I think that this notion of probability should be placed at the end, and as a conclusion, of theoretical considerations, and not as an *a priori* axiom, though I may well admit that this indeterminacy corresponds to experimental possibilities. [...] Must one necessarily elevate indeterminism to a principle?

When Lorentz, in concluding the conference, cautioned against pedantic certainty about the interpretation of quantum effects, he did so with good reason. The reason has been emphasized several times [1]:

> Contrary to popular believe, the interpretation of quantum theory was not settled at this conference, and no consensus was reached.
> There is no longer a definitive, widely accepted interpretation of quantum mechanics; it is no longer clear who was right and who was wrong in October 1927.

In Copenhagen?

In assessing the legacy of the Solvay conference it is pointed out that [1]:

> After 1927, [what is now known as] the Copenhagen interpretation became firmly established. Rival views were marginalised, in particular those represented by de Broglie, Schrödinger and Einstein.
> It seems rather clear that the historical priority of de Broglie's work was being downplayed by Pauli's remarks, as it has been more or less ever since.
> ... the standard (and unbalanced) version of events was propagated in particular by Bohr and Heisenberg, especially through their writings decades later.
> A general opinion arose that the questions had been essentially settled, and that a satisfactory point of view had been arrived at, principally through the work of Bohr and Heisenberg. For subsequent generations of physicists, 'shut up and calculate' emerged as the working rule among the vast majority.
> Despite this atmosphere, the questioning never completely died out ... But attitudes changed very slowly. Younger physicists were strongly discouraged from pursuing such questions. Those who persisted generally had difficult careers ...

Indeed, the arguments had been settled—to the satisfaction of none, on the authority of a few, and ordained as the Copenhagen interpretation. The motivation for codifying the principles of quantum mechanics is explained by Heisenberg [6]:

> SCHRÖDINGER attempted to deny entirely the existence of discrete energy values and quantum jumps, and to resolve quantum theory into a simple classical wave theory.

He then outlines the procedure as follows:

> The months which followed [. . .] were a time of the most intensive work in Copenhagen, from which there finally emerged what is called the "Copenhagen interpretation of quantum theory," [. . .] BOHR intended to work the new simple pictures, obtained by wave mechanics, into the interpretation of the theory, while I for my part attempted to extend the physical significance . . ..

Although the resulting Copenhagen orthodoxy is accepted and respected worldwide it only exists as folklore without formal documentation. The closest to a statement of the assumptions and considerations that were taken into account leading to the final version is to be found in a collection of early reviews by Bohr [7]. As formulated in these, the theory stands on three pillars:

(1) Complementarity
(2) The Correspondence Principle
(3) The Quantum Postulate

Although a precise definition of these concepts is lacking, some important implications are spelled out:

> . . . the essence [of quantum theory] may be expressed in the so-called quantum postulate, which attributes to any atomic process an essential discontinuity, or rather individuality, completely foreign to the classical theories and symbolized by Planck's quantum of action. This postulate implies a renunciation as regards the causal space-time coordination of atomic processes.
> . . . only those quantities which depend on the existence of the stationary states and the possibilities of transitions between them occur in the new theory . . .
> . . . the fundamental postulate of the indivisibility of the quantum of action is itself, from the classical point of view, an irrational element which inevitably requires us to forego a causal mode of description and which, because of the coupling between phenomena and their observation, forces us to adopt a new mode of description designated as *complementary* . . .
> . . . any measurement which aims at tracing the motions of the elementary particles introduces an unavoidable interference with the course of the phenomena and so includes an element of uncertainty which is determined by the magnitude of the quantum of action.
> . . . any observation takes place at the cost of the connection between the past and the future course of phenomena.
> *. . . the finite magnitude of the quantum of action prevents altogether a sharp distinction being made between a phenomenon and the agency by which it is observed . . .*
> . . . wave-mechanical solutions can be visualized only in so far as they can be described with the aid of the concept of free particles.
> We are here so far removed from a causal description that an atom in a stationary state may in general even be said to possess a free choice between various possible transitions to other stationary states. From the very nature of the matter, we can only employ probability considerations to predict the occurrence of the individual processes . . .

The extremes required to formulate a theory that supersedes all else have a distinct smell of desperation. Any theory that discounts any one of: causality, certainty, continuity and the distinction between subject and object, would normally be viewed with scepticism. In addition to these the Copenhagen theory advocates atomic free will, observer-created reality and non-local interaction. The motivation behind this overkill was identified by the philosopher Karl Popper as a morbid fear that Schrödinger's wave mechanics would gain acceptance as the new non-classical paradigm.

To Schrödinger?

Popper [8] dissects the sordid attacks by Bohr, Born and Heisenberg against Schrödinger, that continued for decades. Some of his comments are quoted verbatim:

> The view that theories are *nothing but* instruments or calculating devices, has become fashionable among quantum theorists, owing to the Copenhagen doctrine that quantum theory is *intrinsically ununderstandable* because we can understand only *classical* 'pictures', such as 'particle pictures' or 'wave pictures'. I think this is a mistaken and even vicious doctrine.
> In these discussions, a certain element of irritation or exasperation is sometimes discernible, directed against Schrödinger who, it is indicated, has 'a personal attachment to the waves', and 'is in love with this idea'—quoting Born.
> Both criticize Schrödinger: Born, for relying on the intuitively understandable 3-dimensional waves which result from 'the method of second quantization' ... and by Heisenberg for his 'misunderstanding of the usual' (orthodox) 'interpretation': a misunderstanding which consists in 'overlooking the fact' that in the orthodox interpretation, the 3-dimensional waves of Klein, Jordan and Wigner 'have a continuous density of energy and momentum, like a Maxwell field'. Thus the two defenders of the orthodox view are here seen to agree only in one point: that Schrödinger must be wrong. But their arguments contradict each other.
> Born, so far as I can see does not answer Schrödinger's question (Are there quantum jumps?) at all. Instead, he discusses a completely different question which he introduces in the title of his section 2: *'Are there atoms?'* He claims that Schrödinger's real intention is to deny that atoms exist. This seems to be an inexplicable misinterpretation of Schrödinger's views, and more specifically of what he says in the place criticized by Born. Schrödinger's theory may explain atoms (not as waves, but as wave *structures*); but it does not explain them away.
> While no answer to Schrödinger's question is given by Born, Heisenberg does give an answer. ... Thus his answer to Schrödinger amounts to the assertion that there *are* quantum jumps—exactly as there are quantum jumps in the classical theory.
> These answers strongly suggest that there really appears to be no evidence for quantum jumps. But if this is so, why not say so? And why attack Schrödinger when he says so?
> The answer to the second question—whether the existence of particles is implied in the interpretation of the present quantum theory—of course depends upon the interpretation chosen. Born, Pauli and Heisenberg clearly imply that the existence of particles is part of the interpretation which they accept, which is the orthodox or Copenhagen interpretation, and that they differ in this point from Schrödinger. I find this a little surprising.
> As against this view, I should have thought that Schrödinger asserted with great force the real existence of atoms or particles, even though he tried to explain them as wave structures.

> I did not know that, by saying that a table or chair is made of timber, I should expose myself to the criticism that I have denied the existence of discrete tables and chairs, and that I have done so because I am 'in love' with timber (as Born says of Schrödinger's attitude towards waves).

With all the cards on the table it is with a considerable sense of wonder that one contemplates the uncritical rejection of Schrödinger's views and acceptance of the Copenhagen dogma by 'all competent physicists'. The wild card turned out to be the mathematician John von Neumann with his 'proof' that [9]:

> It is therefore not, as often assumed, a question of reinterpretation of quantum mechanics—the present system of quantum mechanics would have to be objectively false in order that another description of the elementary process than the statistical one be possible.

Although this proof was later falsified by Bell [10] it was concluded more recently that [11]:

> Quantum theory is the deepest explanation known to science ... There is no other.

The irony is that, worse than von Neumann's proof was Bell's 'confirmation' of non-local quantum events, supposed to prove conclusively that quantum theory overrules special relativity. This red herring, to be discussed later, once the essential four-dimensional nature of quantum theory has been demonstrated, managed to restore the blind faith in Copenhagen.

Before the War?

The turbulent decade after Brussels saw a minor diaspora of the important role players and in Popper's words:

> ... chaos ruled in the Copenhagen camp.

Einstein, Fermi and Pauli relocated to the USA, Born settled in England and Schrödinger ended up in Ireland. Apart from von Neumann's analysis the major development in quantum theory was the publication of what became known as *Dirac's equation* and the *EPR Paradox*.

Despite Heisenberg and von Neumann's claims of completeness for quantum mechanics as a theory the single most important non-classical feature, electron spin, had to be added empirically, without proof or understanding. This was achieved by Pauli [12] on modification of Schrödinger's equation. The trick was to add the two-fold variability in electronic motion by writing the wave equation in a magnetic field on a two-component column vector with coefficients in matrix form.

A more serious problem existed in the fact that the wave equation is not covariant, which means that it lacks the generality of relativity theory. The problem lies therein that the time coordinate appears as a first-order differential, compared to the second-order form of the space coordinates. Relativity requires that all space-time coordinates be treated equivalently.

Two different approaches to formulate a covariant matter-wave equation were proposed. Various authors, including Schrödinger, Klein and Gordon, achieved covariance by modifying the general 3D wave equation according to de Broglie's model, in what is generally known as the Klein-Gordon equation. This is, without any doubt the correct way to go, but by assuming a 3D plane-wave solution the covariance is destroyed on separation of the time and space variables.

As an alternative Dirac explored the possibility of linearizing the space part of the equation to match the time variable. This requires writing of the square root of the relativistic energy in linear form:

$$E = \pm\sqrt{(cp_x)^2 + m_0^2c^4} = \alpha(cp_x) + \beta(m_0c^2)$$

Although this is algebraically impossible with α and β in the form of ordinary numbers, it could be achieved by writing the coefficients in matrix form; and as it turned out the required matrices were identical with the Pauli spin matrices.

This result, often hailed as the highest achievement of twentieth-century science, appeared to have solved the covariance and spin problems in one fell swoop. Most textbooks still refer to spin as a relativistic effect. However, it has since been demonstrated that linearization of the squared operator in Schrödinger's non-relativistic amplitude (time-independent) equation reduces to an equivalent form in terms of Pauli matrices. Dirac's equation does not provide a fundamental explanation of electron spin, as often claimed. It empirically adds suitable matrices to a covariant form of the wave equation to reflect the two-fold variability ascribed to spin.

The euphoria that hailed Dirac's achievement was not shared by all contemporary physicists. Heisenberg is quoted to have written [13]:

> The saddest chapter of modern physics is and remains the Dirac theory.

The stumbling block was the fact that Dirac's theory (like the Klein-Gordon!) predicted equal numbers of positive and negative energy states. As it turned out, rather than being its weakest point it was the strongest—eventually leading to the identification of positrons (negative energy electrons) in cosmic radiation.

As in the case of KG (Klein-Gordon), plane-wave solutions are assumed for Dirac's equation, which provides an improved description of angular-momentum and spin states. It was noted [14] however, that Dirac theory leads to absurdities when electrons are located with a precision less than the Compton wavelength, although the probability interpretation demands an electronic radius of zero. This contradiction, known as Klein's paradox could be partially overcome by a field, rather than particle, interpretation of Dirac's equation.

Klein's paradox, being circumvented by turning to a field theory, returned with a vengence as the self-energy of electrons, and other infinities. The problem arises on assuming an electron to be the source of an electric field, and when on the move it interacts with its own field by an inverse-square interaction at zero distance. In the limit, as r tends to zero, the interaction $1/r^2 \to \infty$ tends to infinity.

Alternatively, an electron as point particle is assumed to come into existence by assembly of its charge components from infinity. The work required to assemble an isotropic point particle can be calculately precisely from Coulomb's law, and again, turns out to be infinite.

Classically the problem was overcome by Lorentz on showing that the singularity is avoided by restricting the minimum size of an electron to its *classical radius* of $r_0 = e^2/mc^2$. In Copenhagen quantum mechanics that deals with probability densities this assumption is not allowed. The product of one-particle probability densities in two different points equals a δ-function that peaks in one place only and is equal to zero everywhere else. This argument rigorously defines the electron as a point particle.

The Copenhagen pioneers combined their efforts to avoid the appearance of infinities in developing a quantum field theory [15], starting from Dirac's equation, hole theory, discovery of the positron and the annihilation of electron-positron pairs. Serious differences of opinion are reflected by a series of wild suggestions from all sides [16] and some angry comments, such as Dirac, declaring [17]:

> The only important fact [of existing theory] that we have to give up is quantum electrodynamics ... we may give it up without regrets—in fact because of its extreme complexity, most physicists will be glad to see the end of it.

or Pauli [13]:

> ... it has pleased me that once again I could say something nasty about my old enemy the Dirac theory.

and Heisenberg complaining [13]:

> I regard the Dirac theory ... as learned trash which no one could take seriously.

Even Schrödinger described [18, p. 381] the transcription of the Dirac equation in covariant form as:

> ... nothing but calculational gymnastics ...

As time ran out into the outbreak of war, work on the field theory also petered out, to be continued by a next generation after the war.

To muddy the waters even further Einstein reiterated his arguments, misunderstood in Brussels, by publication [19] of his objections to the non-local nature of Copenhagen theory, in collaboration with colleagues Podolsky and Rosen, widely known as the EPR paradox.

The essence of the argument is grasped most readily in the form reformulated years later by Bohm and by Bell [20] in terms of spin measurements. It invokes the symmetrical splitting of a homonuclear diatomic molecule stabilized by an electron-pair bond. The relative orientation of the unpaired spins, as dictated by the wave function, is preserved as the atoms drift apart, but not in an absolute sense.

According to the Copenhagen measurement principle the spin state on either atom in a given direction is established only when the wave function that rules both atoms is reduced by a magnetic measurement—not only on the atom subject to the measurement, but also at the second atom which is not physically affected by the magnetic probe. The collapsing wave function transmits the outcome of the measurement instantaneously to the second atom, thereby fixing its spin state, irrespective of the orientation of the magnet.

In terms of EPR terminology it is implied that, without superluminal (non-local) interaction, the predicted correlation of independent orthogonal components of spin violates the quantum commutation rules. This paradox ordains a stark choice between quantum and relativity theories. Of all the quantum mysteries this one is the most persistent and no claim about the absolute validity of quantum mechanics can be seriously entertained until the paradox has been resolved.

After the War?

In the war-time political climate there was no longer room for fundamental philosophical study of natural phenomena. All quantum and nuclear scientists were drawn into the technological development of nuclear weapons and radar systems. After the second world war the search for understanding was largely forgotten and financial gain became the driving force. The Copenhagen formula served as a working model, further developed, as needed, through *ad hoc* assumptions. Technological success attracted ever larger financial resources, which stimulated the growth towards Big Science. No longer was there place in the sun for individual genius and to be heard meant speaking with the voice of a possessive sponsor.

It was left to philosophers and fellow travellers with an inadequate grasp of science to ponder the mysteries of quantum theory. It culminated in the system about which is said [11]:

> ... quantum theory—the deepest discovery of the physical sciences—has acquired a reputation of endorsing practically every mystical and occult doctrine ever proposed.

The discussion of Big Science itself will not be pursued any further, but its effect on the status and practice of quantum science cannot be ignored.

The focus of quantum theory had shifted to the USA, as the continuation of Federally funded war-time physics research aimed at the development of sophisticated weaponry. A staggering number of physicists and military dollars were committed to weaponize nuclear fission and radiation sources. A detailed account of the ensuing post-war funding of and approach to science [21] also traces the legacy of this new style of research into the twenty-first century. To quote:

> Anything that smacked of 'interpretation', or worse, 'philosophy', began to carry a taint for many scientists ...
> Openly philosophical areas of physics, the intellectual roots of which stretched back before the war, became increasingly marginalized, such as grand questions about ... the subtle foundations of quantum theory. Sometimes these were denigrated as not even being 'real physics' ...

Because of the military incentive an unprecedented urgency to produce useful results favoured an engineering-style of research, based on:

> ... a gut feeling for the relevant phenomena without getting lost in philosophical niceties.

As a relevant example the development of quantum field theory and quantum electrodynamics is highlighted. We quote:

> ... in 1947, using repurposed microwave-frequency electronics left over from his wartime radar work, physicist Willis Lamb of Columbia University in New York measured a tiny shift ... in the energy levels of an electron in the 2s and 2p orbitals of a hydrogen atom. Lamb's remarkable achievement challenged physicists' prevailing understanding of the vacuum—the mysterious state of lowest-possible energy.
>
> One of the first to hear about the Lamb shift was physicist Julian Schwinger, who before the war had been a rising star in quantum theory. Like so many physicists ..., Schwinger had been forced to rethink his approach to calculation. Elegant derivations from first principles ... were of little value to the many colleagues who needed to fine-tune electronics components for maximum efficiency. Instead ... he internalized from the engineers a modular, 'effective circuit' approach. Rather than calculate the total electrical resistance of a complicated component from the lofty heights of Maxwell's equations, he could 'black-box' each component, substituting its overall resistance as determined from measurements of inputs and outputs.
>
> Schwinger approached the Lamb shift with his ... lessons still fresh. Since the 1930s, senior theorists had tried to calculate the effects of subtle quantum fluctuations from first principles. Maddeningly, their equations always broke down, producing unphysical infinities instead of finite answers. Schwinger arranged his equations in terms of measurable inputs and outputs, just as his engineering colleagues ... had done with real-world electronics. By recasting the calculation, Schwinger managed to calculate the effects of quantum fluctuations on the electron's energy levels and obtain an answer that matched Lamb's measurement to an extraordinary precision.

The same result was obtained by Richard Feynman, a wartime Los Alamos veteran, based on his path-integral model of quantum mechanics. After theoretical consideration of the problem he concluded [22] that:

> Quantum electrodynamics gives nonsensical results.
>
> We do not know whether it is truly quantum electrodynamics or our assumptions about the distribution of charge inside the particles which are at fault. Only when we have a complete theory of these particles and their interactions will we be able to determine the limitations, if any, of our present theory of quantum electrodynamics.

With respect to the infinite term that invalidates the path-integral analysis of the Lamb shift he theorizes:

> ... it was hoped that some day this term would be finite.

As an immediate remedy to ensure urgently required useful results it is decided:

> ... in view of our present ignorance, convergence of the integrals can be made artificially by supplying an additional factor ... in the integrand ... For high values [this factor] cuts off the integral.
>
> All physical quantities are to be calculated by assuming ... this cutoff factor.
>
> If ... the result depends sensitively on [the factor] no quantitative meaning can be given to the result, for the cutoff function is arbitrary and is not completely satisfactory.

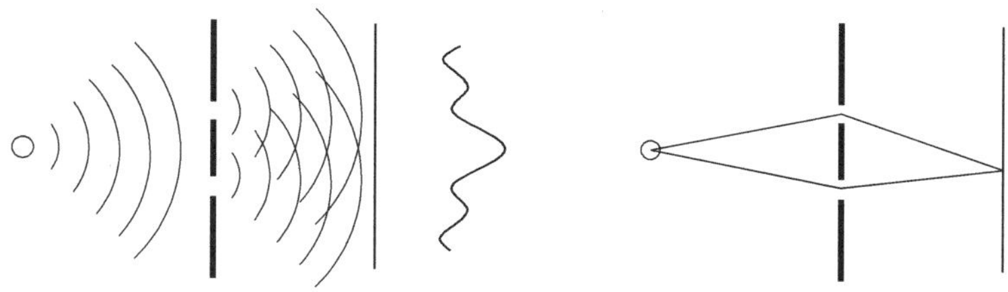

Fig. 3.1 Young's two-slit experiment in which a monochromatic light wave transmitted through two narrow slits produces an interference pattern is shown on the left. Alternative paths for a particle emitted by the same source is shown on the right

For the Lamb shift, said to be insensitive to the cutoff factor, the 'theoretical value is to be trusted'.

In a more recent analysis [23] it is concluded:

> It is quite likely that QED is a low-energy approximation to some complicated high-energy supertheory (a string theory?)
> ... until we discover and solve the supertheory; the couplings must me measured.

Feynman's model was inspired by efforts to explain the interference pattern observed in a two-slit experiment. The solution provided by the wave model, shown in Fig. 3.1, is straight forward and readily demonstrated by adding the wave patterns generated on passage through the slits, at the position of the observation screen.

In Feynman's reconstruction [24] it is argued that a quantum particle has the choice of going through either slit in order to reach a point on the screen. The probability of the particle reaching a given point on the screen, *via* a given slit, is assumed to depend on the length of the path to be followed. The square of the path length, or *amplitude*, defines the probability.

The time it takes for a particle to reach its destination along a path is represented by the final direction of an elementary unit arrow that rotates while the particle moves, starting from a specified zero orientation. In the two-slit experiment there are two possible paths between the source and a point on the screen. Vectorial addition of the two associated elementary arrows then represents the overall amplitude for each point, which squared, specifies the probability of the particle reaching that point.

This simple picture is extended by adding more and more slits to the central screen; generating more and more alternative paths. This requires summation over an increasing number of amplitudes, lying increasingly closer together, until the slits coalesce, the screen disappears and the summation reduces to integration.

To ensure that such integrals produce the same results as the wave model they are defined as action integrals, of the same type as Schrödinger wave functions, over the classical action of the path. As with Schrödinger functions the total action is obtained by a linear combination over all possible paths.

It is immediately obvious that the number of possible paths connecting any two points is infinite. The challenge is to identify those paths of highest probability, on the basis of a suitable physical model. It is argued that in a field-free environment only linear paths, like those shown in Fig. 3.1 need to be considered. However, in a physical vacuum the electron encounters several potential-energy fields, the most important of which is the electromagnetic field. Interaction with the field is responsible for perturbation of the linear paths, creating detours and alternative paths to be taken into consideration. Evaluation of these perturbations requires knowledge of the fields in the vacuum and their nature.

The most important phenomenon predicted by quantum theory is that of zero-point motion, best illustrated by the quantum levels of a harmonic oscillator. It was first postulated by Einstein [25] as a corrective term to Planck's radiation formula, giving rise to the view that the electromagnetic field decouples into an infinite number of harmonic oscillators. Dirac assumed the same model in describing the quantized electromagnetic field as an ensemble of harmonic oscillators. Each oscillator contributes a fixed amount, resulting in infinite zero-point energy in any finite region of the vacuum.

Because of the uncertainty principle zero-point fluctuations are argued to occur all the time, giving rise to short-lived excitations that may have an effect on charged particles, such as electrons.

A typical vacuum polarization event results in the production of "virtual" electron-positron pairs, shown schematically as a "one-loop" event:

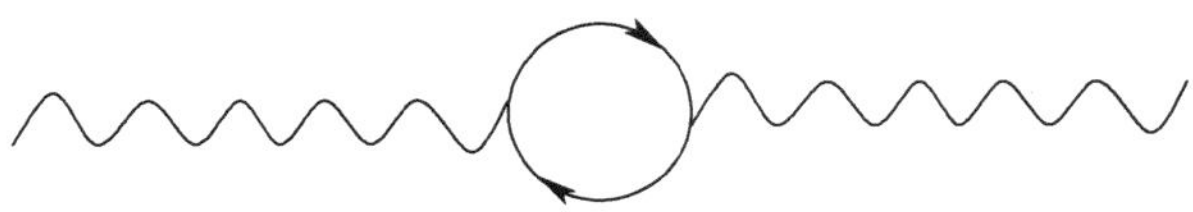

Quantum electrodynamics (QED) mathematically describes all interactions involving electrically charged particles by means of the exchange of photons. In Feynman's model, interaction between electrons and photons is modelled as the interaction between separate electron and photon fields. Symbolically the interaction is formulated in terms of three basic actions:

- a photon that moves between space-time points with amplitude $P(A \rightarrow B)$;
- an electron that moves, amplitude $E(C \rightarrow D)$;
- an electron emits or absorbs a photon, amplitude j.

Feynman diagrams represent the three actions by a straight line for an electron, a wavy line for a photon and the junction of two straight lines and a wavy one as an absorption or emission vertex.

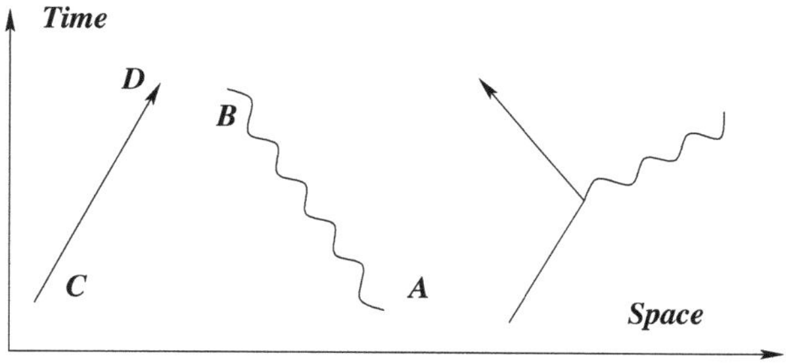

The amplitude for direct motion of an electron depends on its mass and its amplitude for coupling with a photon depends on its charge. However, because of vacuum fluctuations these amplitudes are modified in any real situation, because of 'corrections' such as:

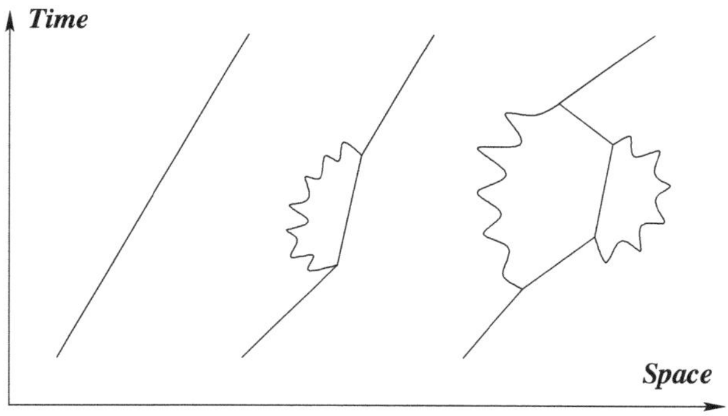

The mass and charge as defined here can obviously never be observed and correspond to a hypothetical 'bare' mass and charge. For amplitudes, fully modified by the vacuum, the correct proportionality factors would be the observed mass and charge of the electrons, but these are not predicted by the theory. What happens in practice is that refinement of the bare values, by the addition of more and more correction factors, due to more closely spaced perturbations, produces a mass and charge that tends to infinity. The effect is the same as that of infinite self energy. Here again, the energy carried by a single electron is not simply the bare value, but that of an electron dressed by a cloud of photons.

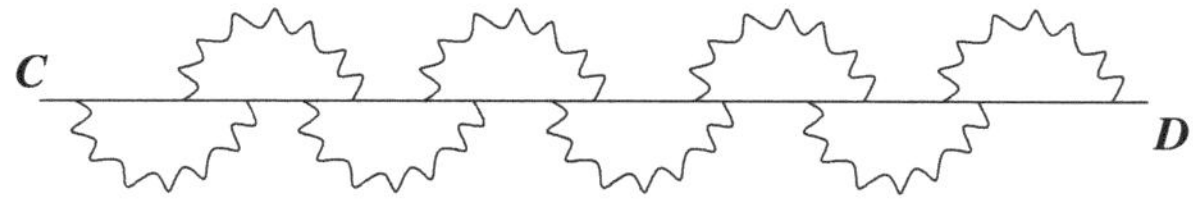

Each additional loop introduces an additional correction and increases the magnitude of the mass and charge parameters beyond their bare values. The secret is to know when to stop adding corrective loops. By excluding the infinite number of perturbations beyond an optimal cutoff must produce the desired result.

This procedure faces two problems. In the first place there is the mathematically dubious procedure of producing a meaningful result as the difference between two infinite quantities. This is the operation that Dirac rejected in principle. The second, equally devastating factor is the inability to determine the correct cutoff.

By trial-and-error it could be established that meaningful results are obtained by assuming the *observed* electron mass as a cutoff. The coupling constant, j, had been determined empirically and, rather than the measured charge, e, of an electron, was found to be given by the spectroscopic fine-structure constant, $\alpha = e^2/\hbar c \simeq 1/137$, in natural units. Although Feynman [24] referred to the value of this constant as "one of the *greatest* damn mysteries of physics ...", it has the logical structure of relating electronic charge to photon velocity as a function of Planck's constant.

The concept of renormalization was also analyzed by Feynman [24]:

> ... no matter how clever the word is, it is what I would call a dippy process! Having to resort to such hocus-pocus has prevented us from proving that the theory of quantum electrodynamics is mathematically self-consistent.
> ... I suspect that renormalization is not mathematically legitimate.
> ... a bunch of words ... is not good mathematics.

He adds in a footnote:

> It has been suggested that ... inconsistencies arise because we haven't taken into account the effects of gravity—which are normally very, very weak, but become important at [very small] distances ...

Despite all of these reservations the 'success' of QED inspired the development of theories, based on the same principles, to account for interactions in atomic nuclei. The broad field of quantum field theory (QFT) that includes quantum chromodynamics (QCD) and grand unified theory (GUT) has been described [26] as 'by far' the most difficult theory in modern physics.

In Chemistry?

The tired statement that "quantum theory explains all of chemistry", has been repeated so often that it now appears routinely as an axiom in many chemistry and popular science texts. To put the record straight we may turn to one of the most respected exponents of chemical theory. Peter Atkins wrote [27]:

> The apotheosis of present-day quantum mechanics seems to be quantum electrodynamics, although it is not entirely clear that the theory exists.

and continues:

> The nature of the chemical bond ... is one of the central topics and successes of the application of quantum theory to chemistry.
> Modern quantum chemistry interprets the [...] rules of thumb [of elementary chemistry] in terms of the quantum-mechanical properties of electrons and nuclei.

These are brave, but meaningless, words. After an extensive review of the situation Atkins concludes (under the entry: **bond**) that:

> ... the conventional view ... is a sweet seduction.
> ... an excellent example of the power of myth in chemistry ...
> ... to understand the shape and stability of a molecule we must study its energy. Why energy is important is of course a much deeper problem.[2]
> A decrease in energy ... comes from a contraction which occurs in the atomic orbitals on each nucleus ...
> These conclusions run ... counter to anything to be found in textbooks.

The author wisely refrains from a quantitative analysis of his conclusions. There is none.

To examine some of the textbook 'myths' we quote from a discussion on 'Atomic Orbital Theory' from a modern popular textbook [28]:

1. The wave character results in a probability interpretation of electronic behavior.
2. In wave mechanics, the wave function is synonymous with an orbital.
3. [Eigenfunction] ... is an acceptable solution of the wave function, which can be an orbital.
4. An exact solution of the Schrödinger equation ... for the hydrogen atom ... may be distinguished by a set of three quantum numbers, n, l, and m ...
5. This quantum number [l] does not enter into the expression for the energy of an orbital.
6. When $n = 2$ and $l = 1$ (the p orbitals), m_l may ... have values of +1, 0, −1, corresponding to the three $2p$ orbitals.
7. ... for each value of the principal quantum number n (except for $n = 1$), there will be three p orbitals corresponding to $m_l = +1, 0, -1$. In a useful convention, these three orbitals, which are mutually perpendicular to each other, are oriented along the three Cartesian coordinate axes and are therefore designated as p_x, p_y, and p_z.
8. According to this [exclusion] principle ... a maximum of two electrons can occupy an orbital, and then, only if the spins of the electrons are opposite (paired), that is, if one electron has $m_s = +1/2$, the other must have $m_s = -1/2$. Stated alternatively, no two electrons in the same atom can have the same values of n, l, m_l and m_s.
9. According to this [Hund's] rule ... a single electron is placed in all orbitals of equal energy (degenerate orbitals) before a second electron is placed in any one

[2] The implication is that the issue is too arcane to be understood by the uninitiated.

of the degenerate set. Furthermore, each of these electrons in the degenerate orbitals has the same (unpaired) spin.
10. The B group elements are the *transition metal elements*; these are the elements with electrons in partially filled $(n-1)d\ldots$ orbitals.
11. The mathematical mixing of two or more different orbitals on a given atom to give the same number of new orbitals, each of which has some of the character of the original component orbitals. (sic) Hybridization requires that the atomic orbitals to be mixed are similar in energy. The resulting hybrid orbitals have directional character, and when used to bond with atomic orbitals of other atoms, they help to determine the shape of the molecule formed.

These statements are all wrong; or, in a few cases, no more than clumsy half-truths. Others are so delusional as to be, in the words of Pauli, 'not even wrong'.

It is one big charade based on the absurd redefinition of a wave function as an orbital, with a set of mysterious properties. Wave functions are solutions of a differential equation, typically a complex exponential such as

$$\Phi(\varphi) = (1/\sqrt{2\pi})\exp(im_l\varphi)\,, \text{ where } i = \sqrt{-1}.$$

The three functions with $m_l = +1, 0, -1$ represent the degenerate set of p functions, two of which are complex and only one (with $m_l = 0$) is real: $\exp 0 = 1$.

This result directly invalidates the statements numbered 2, 3, 5, 6 and 7. The major delusion arises from the assumption of three degenerate **real** wave functions, defined as the p_x, p_y and p_z orbitals. Each of these must have, by definition, the quantum number $m_l = 0$, in conflict with the exclusion principle and Hund's rule as stated in 8 and 9.

For the rest, statement 1 is just poppycock and 4 confuses a hydrogen atom with an electron. Statement 10 is a deliberate falsehood as the so-called B group of elements, as shown on any version of the periodic table, includes groups IB and IIB, with filled d shells. Statement 11 contradicts itself in that the s and p levels which are hybridized for the carbon atom are not degenerate.

The problem here is that the outline on atomic orbitals [28] is not a harmless simplification for beginners,[3] but it also features as the basis of *ab initio* linear-combination-of atomic orbitals (LCAO) molecular-orbital (MO) computational chemistry—the flagship of quantum chemistry. Orbital hybridization is not mentioned explicitly but is skillfully smuggled in by assuming real atomic orbitals as the geometrical elements in setting up the MO's by group theory. It is re-emphasized that real orbitals have no quantum-mechanical meaning and the way they are handled here by computational methods that resemble quantum formalism is wasted effort and unnecessary window dressing.

These methods of quantum chemistry resemble QED in a sinister way. Instead of using an infinite basis set as claimed in theory, a very limited selection is made

[3]The Wikipedia entry on *atomic orbital* regurgitates the same material.

in practice and the computation is terminated as soon as a desired result, such as an experimentally known molecular structure,[4] is reproduced. This is like assuming a cutoff in QED dictated by the measured values of electronic mass, charge and classical radius.

The orbital concept, like renormalization, rocketed into prominence after the war, championed by Pauling in the USA and Coulson in Britain, uncritically based on Copenhagen quantum theory. Emboldened by Pauling's irresponsible statement that [29]:

> ... if quantum theory had been developed by the chemist rather than the spectroscopist it is probable that the tetrahedral orbitals ... would play the fundamental role in the theory, in place of the *s* and *p* orbitals

the chemist jumped to the conclusion that the quantum numbers of real *p* orbitals can be ignored with impunity, and never looked back. Today the orbital myth is more deeply ingrained in the psyche of the chemist than renormalization in the mind of the spectroscopist. The misguided belief that all of chemistry is miraculously predicted by quantum theory is eclipsed only by the bizarre notion of quantum cosmology, most likely inspired by the profound insight that [30]:

> In quantum mechanics almost everything is uncertain. So why should the topology of space-time be fixed?

The mind boggles.

In Cosmology?

The traditional view of quantum theory considered it confined to micro systems where Planck's constant has a measurable effect on dynamic processes. In macro systems where h is insignificant it merges into classical theory. Big-bang cosmology that postulates the birth of the universe in a singular point accordingly assumes this event defined by a universal wave function, which still drives a cosmic expansion. In this theory there is no room for the conventional classical limit where the relative value of Planck's constant approaches zero. It is replaced by a more general definition [31]:

> If quantum mechanics is the underlying framework of the laws of physics, then there must be a description of the universe as a whole and everything in it in quantum-mechanical terms.

The rationale behind this is the lack of an external observer to collapse the wave function in the way specified by the Copenhagen interpretation. But the existence of an observable classical reality presents a new dilemma: What did the universe

[4]To kickstart the computation it is necessary to assume an experimentally known molecular structure.

consist of before the emergence of the first intelligent observer to create an objective reality? How did this primary observer materialize from a collapsing wave function, and where did he get a PhD from [32]? Why did this first 'measurement' not arrest the cosmic expansion?

To get around the measurement problem quantum cosmologists have popularized the concept of quantum decoherence, introduced by Zeh [33]. The idea harks back to Dirac's 1927 Solvay argument, in dispute with Heisenberg's measurement principle, that an observed eigenvalue "must be regarded as chosen by nature", and to Everett's 'many-world' theory.[5] The aim is to determine which Feynman histories have appropriate probabilities for course-grained alternatives to decohere[6] as if a measurement had been performed.

In that the collapse of a wave function corresponds to an irreducible irreversibility, decoherence is ascribed to the interaction of the environment with a quantum system. In summary [31]:

> Decoherence ... generalizes and replaces the notion of "measurement", which served this role in the Copenhagen interpretations.

leading to the breathtaking conclusion that:

> ... resolution of the problem of interpretation presented by quantum mechanics is not to be accomplished by further intense scrutiny of the subject as it applies to reproducible laboratory situations, but rather through an examination of the origin of the universe and its subsequent history. Quantum mechanics is best and most fundamentally understood in the context of quantum cosmology.

In response to the presentation [31] the following exchange is reported:

J.-P. Vigier: The problem with this way of presenting the problems of the history of the Universe since its origin is that we do not ever know (to quote Voltaire) that the Universe was created or not, i.e., it might have infinite history. The various questions discussed in present day cosmology, the redshift controversy, the physical composition and distribution of the matter and of its dominant forms (plasma?) and the existence or not of a "big bang", are not introduced on this type of models. Unless those it is difficult to distinguish scientific speculation from science fiction.

J.B. Hartle: Thank you for your comment.

The erudite response to Vigier's remark is anything but convincing reassurance that all of this is not science fiction. It borders on the arrogant to demand the reformulation of quantum theory, despite its deficiencies, to be based on the big-bang model. However, given the well-known research interest of the first author it is not so much the interpretation of quantum theory that is at issue, than the assumptions of elementary-particle physics.

---

[5] In which a measurement splits the history of the universe into two branches, only one of which remains in evidence.

[6] Meaning that the superposition principle no longer applies.

In Particle Physics?

Modern particle physics has the problem of operating without an experimental basis. Its main pursuit is to formulate a grand unified theory that includes all possible fields. Its standard technique is the renormalization of energy models under broken internal, or gauge symmetries. The most powerful particle accelerators do not operate at sufficiently high energies for direct study of the symmetry breaking that generates separate gravitational, electroweak and strong nuclear forces, but assuming that this happened soon after the big bang, cosmology may provide the missing data to substantiate such conjecture.

There is clearly no room for doubt. Without a big bang, particle physics has no leg to stand on and therefore Vigier's doubts are too gastly to contemplate. In addition, the more sophisticated inflationary big bang supplies the ready-made model of symmetry breaking in the form of a vacuum phase transition, free of charge. The kickback is that the Higgs particle of physics confirms the existence of a field, which is required to generate mass from cold dark matter. The purported observation of a Higgs particle is now claimed to provide irrefutable proof of the big bang. The great oracle of big-bang particle theory, Stephen Hawking, provides the ultimate elucidation [34]:

> ... the history of a particle is now a complete curved space-time that represents the history of the whole universe [...] these curved space-times must be taken Euclidean. [...] if we knew how the Euclidean curved space-times [...] behaved at early times, we would know the quantum state of the universe.

Surely this cannot be serious science. Not only is Euclidean curved space an oxymoron, but if this is quantum theory, Feynman was right when he said that nobody understands quantum theory.

Although symbiosis between particle physics and cosmology is complicated by the cosmological constant, which is close to zero in cosmology and approaches infinity with the vacuum particles of QFT, this discrepancy is conveniently glossed over as 'quantum uncertainty'.

To Quantum Theory?

The confusion over quantum theory is largely due to the way in which the articles of Copenhagen orthodoxy were gradually overturned over a period of fifty years, without retraction of the original formulation. Bohr's formulation of the theory in terms of a quantum postulate, complementarity and a correspondence principle was too vague to have operational meaning and its impact was minimal. A working model that developed from these notions were based on the superposition principle and the particle-wave model. Superposition implied uncertainty and the particle-wave only made sense as a probability construct. The probabilistic concept is no more than a crude effort to simulate wave motion by the random motion of a point particle.

Nowhere has the superposition principle been abused more systematically than in quantum chemistry, where it serves to convert complex into real wave functions, creating orbitals. Although orbitals have no quantum numbers or meaning they are used, in turn, to make more superpositions, not allowed for classical systems. It is remarkable how these messy objects even managed to penetrate into solid-state and molecular physics.

Whenever the Copenhagen axioms experienced interpretational problems they were augmented by auxiliary assumptions, such as the collapse of (an undefined) wave packet to rationalize the effects of quantum measurement, giving a new twist to the probability concept. A logical outflow of the measurement postulate was the many-worlds interpretation and redesign of probability as the sum over histories. This led to the reformulation of the quantum mechanics of a particle into a field theory, in conflict with the fundamental concept of quantum jumps.

By now the noble ideas of Heisenberg and Born to develop a theory only on the basis of experimental evidence, without reference to space or time, were all but forgotten. The only firm conviction that remained was that Schrödinger's wave model had to be resisted at all costs. Schrödinger's re-interpretation [35] of Weyl's gauge principle as a phase change, strictly a wave concept, was next incorporated into particle physics as the property of a point particle. At this stage it became necessary to establish concordance with quantum gravity and the cosmic origin at a space-time singularity.

To avoid the embarrassing absence of an observer to create objective reality it was necessary to replace the measurement postulate with decoherence, in conflict with the superposition principle. Although Bohr's model had disappeared without a trace, it remains in vogue!

In its original form the concept of quantum measurement and uncertainty is a precise analog of a roulette wheel. The uncertain discrete position at which the pointer settles, correlates with the way in which the gamble of a quantum measurement pays off in a randomly selected eigenvalue. In the extended quantum theory this option is no longer on the table and it is fair to conclude that the quantum gamble has failed.

And that is not the end of it. Quantum cosmology proudly claims to be consistent with the theory of general relativity. This is amazing in view of the generally accepted non-local nature of quantum theory, which conflicts directly with the theory of special relativity.

We conclude that quantum theory must rank as one of the most disastrous public-relations exercises of the twentieth century and makes up a prominent part of what has been described [36] as fairy-tale physics.

The final quantum gamble is playing out in what is known as quantum computing. Whereas 'classical' computing relies on the manipulation of binary digits, or bits, such as 0 and 1, a quantum computer manipulates *qubits* that arise from the 'quantum superposition' and entanglement of classical bits. Quantum computers are generally believed to be superior and faster, or at least as fast, as classical computers for any computing task. There is no proof of that and it has never been demonstrated in practice. Even the ability of quantum computers to decrypt many of the cryptographic systems in use today only exists in theory.

In theory, quantum computing relies on the superposition of binary digits and their entanglement to enable free flow of information. Both of these attributes are strictly confined to linear systems and therefore to be realized only at absolute zero temperature, whatever the working medium. The most promising candidate at present is the hypothetical two-dimensional particle called an *anyon* that only 'exists' in Euclidean space. It is doomed to evaporate in curved space-time.

References

1. G. Bacciagaluppi, A Valentini, *Quantum Theory at the Crossroads: Reconsidering the 1927 Solvay Conference* (Univ. Press, Cambridge, 2009)
2. M. Jammer, *The Conceptual Development of Quantum Mechanics* (McGraw-Hill, New York, 1966)
3. M. Jammer, *The Philosophy of Quantum Mechanics: The Interpretations of Quantum Mechanics in Historical Perspective* (Wiley, New York, 1974)
4. E. Schrödinger, Zur Einsteinschen Gastheorie. Physik. Zeit. **27**, 95–101 (1926)
5. Howard, in *Sixty-two Years of Uncertainty*, ed. by A.I. Millar (Plenum, New York, 1990)
6. W. Heisenberg, *The Development of the Interpretation of Quantum Theory*, in *Niels Bohr and the Development of Physics*, ed. by W. Pauli (Pergamon, London, 1955)
7. N. Bohr, *Atomic Theory and the Description of Nature* (Cambridge University Press, London, 1934)
8. K.R. Popper, *Quantum Theory and the Schism in Physics*, (1982). Reprinted (Routledge, London, 1995)
9. J. von Neumann, *Mathematical Foundations of Quantum Mechanics*, (1932). English translation (Princeton University Press, Princeton, 1955)
10. J.S. Bell, On the problem of hidden variables in quantum mechanics. Rev. Mod. Phys. **38**, 447–452 (1966)
11. D. Deutsch, *The Beginning of Infinity* (Viking, New York, 2011)
12. W. Pauli, Zur Quantenmechanik des magnetischen Elektrons. Z. Phys. **43**, 601–623 (1927)
13. A. Hermann, K. von Meyer, V. Weisskopf (eds.), *Wolfgang Pauli, Scientific Correspondence* (Springer, New York, 1975)
14. O. Klein, Die Reflexion von Elektronen an einem Potentialsprung nach der relativistischen Dynamik von Dirac. Zeitschrift für Physik **53**, 157 (1929)
15. C. Itzykson, J.-B. Zuber, *Quantum Field Theory* (McGraw-Hill, New York, 1985)
16. A. Pais, *Inward Bound* (Univ. Press, Oxford, 1986)
17. P.A.M. Dirac, Does conservation of energy hold in atomic processes? Nature **137**, 298–299 (1936)
18. A. Rüger, Atomism from cosmology. Stud. Hist. Phil. Sci. **18**, 377–401 (1988)
19. A. Einstein, N. Rosen, B. Podolsky, Can quantum-mechanical description of physical reality be considered complete? Phys. Rev. **47**, 777–780 (1935)
20. J.S. Bell, On the Einstein–Podolsky–Rosen paradox. Physics **1**, 195–200 (1964)
21. D. Kaiser, History: Shut up and calculate! Nature, **505**, 153–155 (2014)
22. R.P. Feynman, A.R. Hibbs, *Quantum Mechanics and Path Integrals*. Emended by D.F. Styer (Dover, New York, 2005)
23. G.P. Lepage, *What is Renormalization?* arXiv:hep-hp/0506330v1 30 Jun (2005)
24. R.P. Feynman, *QED* (Penguin, Harmondsworth, 1990)

25. A. Einstein, O. Stern, Einige Argumente für die Annahme einer molekularen Agitation beim absolute Nullpunkt. Annalen der Physik **40**, 551 (1913)
26. E. Whitten, *Beautiful Minds*, vol. 20 (LaRepublica, 2011)
27. P.W. Atkins, *Quanta*, Oxford Chemistry Series 21 (Clarendon Press, Oxford, 1974)
28. M. Orchin, R.S. Macomber, A. Pinhas, R.M. Wilson, *The Vocabulary of Organic Chemistry*, 2nd edn. (Wiley, 2005)
29. L. Pauling, *The Nature of the Chemical Bond*, 3rd edn. (Cornell Univ. Press, Ithaca, 1960)
30. B. Schwarzschild, Why is the cosmological constant so very small? Physics Today, 21–24 (1989)
31. M. Gell-Mann, J.B. Hartle, Quantum mechanics in the light of quantum cosmology. In *Proc. 3rd. Int. Symp. Foundations of Quantum Mechanics*, Tokyo, pp. 321–343, 1989
32. J.S. Bell, *Quantum Mechanics for Cosmologists* in *Speakable and Unspeakable in Quantum Mechanics* (Univ. Press, Cambridge, 1993)
33. H.-D. Zeh, On the interpretation of measurement in quantum theory. Found. Phys. **1**, 69–76 (1970)
34. S.W. Hawking, *A Brief History of Time* (Bantam, London, 1988)
35. E. Schrödinger, Über eine bemerkenswerte Eigenschaft der Quantenbahnen eines einzelnen Elektrons. Z. für Physik **12**, 13–23 (1922)
36. J. Baggott, *Farewell to Reality* (Pegasus Books, New York, 2013)

Chapter 4
The Classical Basis

The first centuries after Newton saw spectacular progress in the application and refinement of mechanics as a fundamental science. It survived the introduction of modifications to deal with special cases, but, in the form of classical mechanics that deals with common situations it remained intact. This is not the case with quantum mechanics that, hundred years on, still struggles to find a generally accepted formulation.

The history of twentieth century science reveals, not only the emergence of powerful concepts that inspired unprecedented technological developments, but also increasing confusion over the philosophical implications of the seminal theories. The major source of confusion, and of serious frustration, occurs in the blatant incompatibility of the two basic theories of physics. The fundamental concept that underpins the theory of relativity is a physical limitation imposed on the maximum speed attainable in nature. In stark contradiction quantum theory is fundamentally non-local; meaning that correlated events remain in instantaneous interaction, irrespective of spatial separation.

Although most scientists have managed to avoid the logical implication of this dilemma by using, and even combining, the two theories, a cold understanding of Nature is impossible while the confusion lasts.

The problem is aggravated by another schizophrenic situation within quantum theory itself. The conceptual difference between atomistic and holistic worlds are ignored by the quantum-mechanical concept of wave-particle duality, with its spin-offs of quantum uncertainty and probabilities.

The dilemma has been debated for a century without success. With the original protagonists no longer around it may now be useful to re-examine the data and the assumptions behind the two theories, for common ground. The new theories were formulated when the classical theories of physics, based on Newtonian mechanics, failed to account for important experimental observations related to electromagnetic and optical phenomena, not directly addressed by the laws of mechanics.

J.C.A. Boeyens, *The Quantum Gamble*, DOI 10.1007/978-3-319-41621-2_4

The recognition of an electromagnetic field by Maxwell rivals the postulate of the law of gravity in scientific importance. The field concept avoids the problem of action at a distance. A spectacular demonstration of a field is provided by sprinkling iron filings on a sheet of paper that covers a bar magnet. They settle into a pattern which is interpreted as magnetic lines of force. This idea was established by Michael Faraday (1791–1867) who demonstrated that electric currents have the same effect on iron filings as permanent magnets and inferred that magnets should have a reciprocal effect on electric currents.

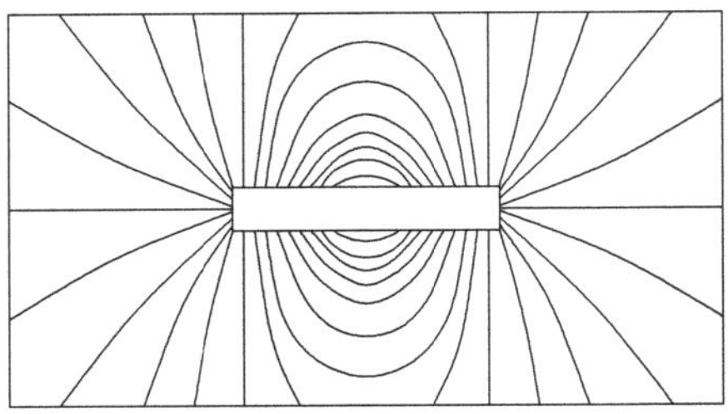

This reciprocal relationship, known as electromagnetic induction, provided the experimental basis on which Maxwell managed to derive a mathematical definition of the electromagnetic field, that requires no material substrate. Faraday remained convinced that the gravitational interaction must also arise from a field, which he tried to demonstrate. Although he failed to do so, he concluded the account of his experiments with the statement (1850):

> The results are negative. They do not shake my strong feeling of the existence of a relation between gravity and electricity, though they give no proof that such a relation exists.

One basic idea of a field is that entities such as electric charges and magnetic poles (or gravitational mass points), known as sources of the field, cause distortion of space in their immediate vicinity, spreading like the lines of force shown in the diagram above. Implicit in this assumption is that space is more than a void, better referred to as aether. It could be shown experimentally that distortion of the aether to establish a field, even in a vacuum, encounters some resistance. In the case of a magnetic field the resistance is measured as *permeability* (μ_0) and for electric fields, as *permittivity* (ϵ_0).

These properties of free space imply a structured medium, and it is not obvious why this conclusion has been said to be ruled out by the theory of special relativity. In fact the existence of the aether is not categorically rejected as only the mechanical properties assumed for the aether, like those postulated by Descartes, or as a medium that represents an absolute frame of reference, become superfluous in relativity theory. Under another name—space—it became the carrier of all fields, implicated in mediating action at a distance.

Maxwell did not have the same problem and freely dealt with vortices in the aether to rationalize his ideas on the electromagnetic field. To him all space was occupied by an aether capable of electric polarization. It meant that electric flow was not restricted to only occur in conductors, but could happen in space. This observation eventually led to Maxwell's detailed theory of electromagnetic radiation.

As is often the case, several other scientists were investigating related phenomena, such as the speed of signals along electric cables, which consistently appeared related to a product of the constants μ_0 and ϵ_0. Maxwell interpreted this to mean that "light consists in transverse undulations of the same medium which is the cause of electric and magnetic phenomena". By this theory light is one form of electromagnetic radiation propagated at the constant speed, $c = 1/\sqrt{\mu_0\epsilon_0}$, in the vacuum. Confirmation of the theory came with the experimental production and transmission of electromagnetic (radio) waves that obey Maxwell's wave equation, by Heinrich Hertz (1857–94) in Germany.

To appreciate the full impact of Maxwell's characterization of the electromagnetic field some background knowledge on the nature of waves and wave motion is essential.

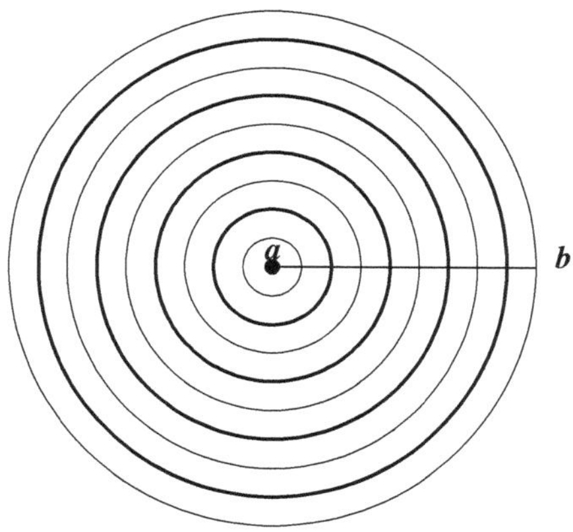

Most, if not all, readers should be familiar with waves on water surfaces. By dropping a small object into a quiet pond the disturbance on the surface is seen to spread in the form of concentric circles. A more general wave pattern is obtained by touching the surface of a mercury pool with the vibrating tip of a tuning fork. The dark circles in the diagram represent the crests of the spreading waves and the fainter ones the troughs. In cross section (*ab*) the wave pattern resembles a sine curve.

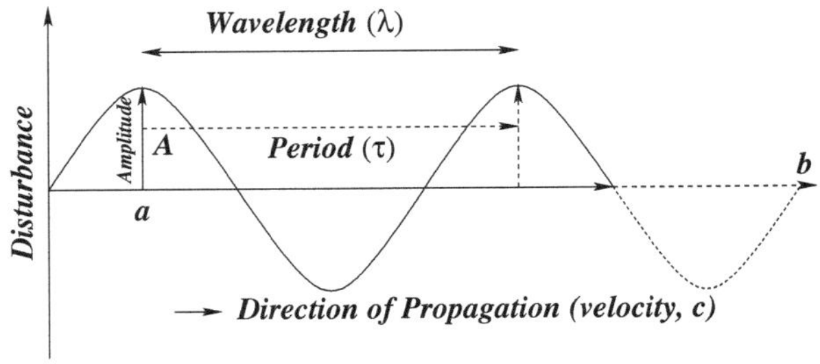

A leaf in the path of a wave that spreads on a pond is observed not to travel with the wave but to move periodically up and down with the passing wave crests and troughs. The time taken by the leaf to complete one cycle is known as the period (τ) of the wave. The distance between successive crests is called the wavelength (λ). Other parameters often used to describe a wave include the amplitude (A), the frequency $\nu = 1/\tau$, wave number $\bar{\nu} = 1/\lambda$ and the speed $c = \lambda/\tau$.

Electromagnetic waves differ from surface waves as they spread in three dimensions and in consisting of two components, an electric and a magnetic component. The disturbance, measured in any direction perpendicular to a radial line (*ab* say) again has the form of a sine curve and the same wave parameters as before, are used to describe both electric (E) and magnetic (H) vibrations. As shown by Maxwell's analysis any electric vibration is always linked to a perpendicular magnetic vibration, both at right angles to the direction of propagation ($E \times H$), as shown schematically below.

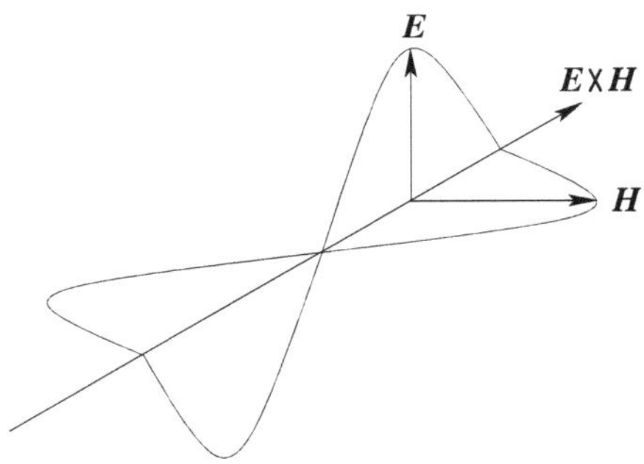

The situation pictured in the diagram is very special in the sense that all electric vibrations occur in a single plane. A light beam with this property is known as plane polarized light. It occurs when light is passed through certain crystals, such as Iceland spar, with a special vibration direction that promotes optimal transmission of a light beam. All electric, and magnetic, vibrations line themselves up to follow this direction of easy passage. The result is that the incoming beam is split into two: an ordinary ray with electric vibrations in the special direction and an extraordinary ray with magnetic vibrations in this direction. These two beams move through the crystal with different velocities. Viewing an object through a crystal of Iceland spar therefore creates two images. Certain synthetic materials, such as herapathite, has the property of not only polarizing a light beam, but also to absorb one of the rays and transmit pure plane polarized light. Polaroid devices are manufactured from such materials.

The wave theory of light demonstrated that all forms of radiation have common characteristics, such as speed of propagation through the vacuum, except for wavelength or frequency. The range of wavelengths that differentiate between different forms of radiation is known as the electromagnetic spectrum. Arranged from short to long wavelength (high to low frequency) the following common types are distinguished:

γ-ray	X-ray	ultraviolet	visible	infrared	microwave	radio

Visible light is further subdivided, by colour, that varies with wavelength, into:

Violet	Indigo	Blue	Green	Yellow	Orange	red

When the colours in sunlight are separated by scattering on water droplets, such a spectrum is observed as a rainbow. When the separation is performed with optical instruments in the laboratory the colour spectrum has a number of black lines, known as Fraunhofer lines, superimposed at some sharply defined wavelengths.

The simplest way to demonstrate the colour composition of white light is by passage through a glass prism which separates the rainbow colours as the beam spreads out.

The relationship between the continuous, Fraunhofer and emission line spectra is shown schematically in the diagram. Fraunhofer lines are due to absorption in a cool gas cloud that would produce its characteristic lines at the same wavelength, when hot.

Recognition of the electromagnetic spectrum opened up a new world of research to study the interaction between radiation and matter, especially chemically pure substances, with enormous implications on the understanding and composition of the cosmos. It was soon noticed by a new breed of scientists, known as spectroscopists, that different chemical elements in an incandescent state, at high temperatures, produced radiation characteristic of each element. These emission spectra of pure elements are not as continuous as the solar spectrum, but consist of characteristic lines at sharply defined wavelengths, exactly like the Fraunhofer lines.

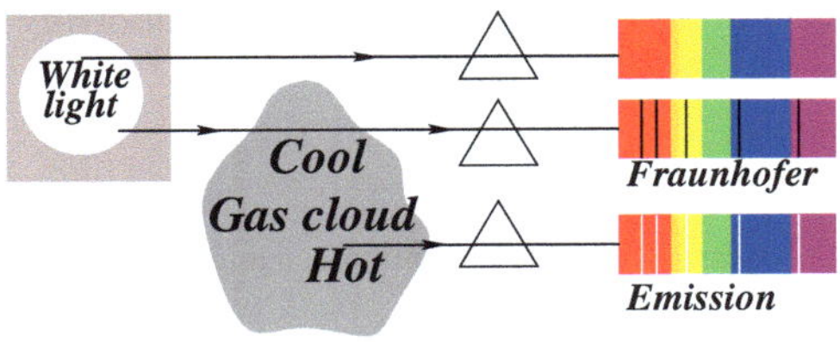

The observation that some Fraunhofer lines occurred at the same wavelengths as the emission lines of common elements immediately enabled the spectrochemical analysis of the atmosphere of the sun and other stars. It is argued that the continuous solar spectrum of light is emitted from the hot interior of the star and that the atoms in the cooler chromosphere that surrounds the sun absorb those frequencies that match their own characteristic emission lines.

The pioneering work on emission spectra was done by Robert Bunsen (1811–99) and Gustav Kirchhoff (1824–87) in Heidelberg. The nature of line spectra remained a mystery for many years and only deepened when Balmer noticed that the spacing of the lines could be described by a simple numerical formula. The formula, coming as it did, from a secondary school teacher and not from an established physicist, was treated as a mere curiosity for many years. It took 28 years (15 years after his death) before the fundamental importance of the Balmer formula was formally recognized.

The full impact of Maxwell's wave equation would be felt early in the new century as it led directly to the development of quantum mechanics and the relativity principle, the two monumental twentieth century theories of physics.

Although Maxwell's electromagnetic model produced several useful new results it appeared to contradict existing physics in several important respects. The wavelength distribution of radiation in a closed cavity, also referred to as a black body, was, for instance, found in blatant contradiction of statistical thermodynamics, whilst the constant speed of light violates the kinematic rules for objects in relative motion. Two individuals are credited in retrospect with the resolution of these dilemmas, but as noticed before for other related situations, most physicists of the world were engaged with these problems and steadily contributed to their solution.

First to crack was the black-body problem.

The term *black body* is a bit of a misnomer, as most black bodies are not black at all. The essential property of what is called a black body is its ability to absorb radiation

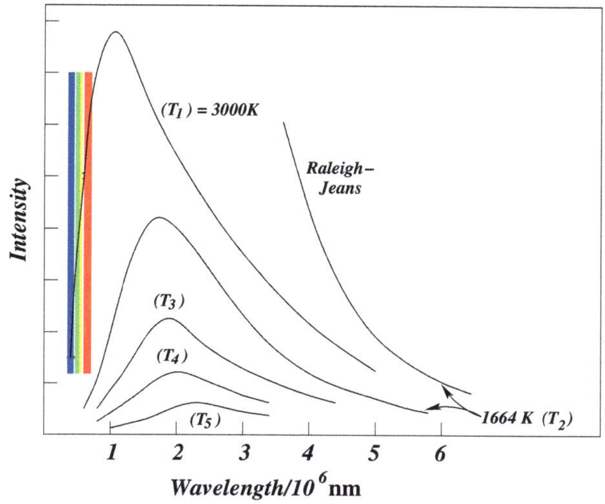

of all wavelengths that falls onto it, without reflection. All stars, including the sun, are almost perfect black bodies. The most convenient device to study the spectrum emitted by a black body is a closed canister that radiates through a small hole when heated.

To anticipate the results we recall that as the temperature of a furnace or even a metal bar increases it glows with increased intensity—dull red at first, turning brighter and eventually reaching white heat as all colours of the spectrum are compressed together. As the amount of radiated energy increases the dominant wavelength goes down, as the frequency goes up.

The spectrum of black-body radiation at various temperatures ($T_1 > T_2 > \cdots > T_5$) is shown graphically above, together with a theoretical curve, based on classical thermodynamics—as calculated by Raleigh and Jeans. The theoretical curve is based on the classical concept of energy equipartitioning.

The Raleigh–Jeans calculation assumed that, at constant temperature, the radiation inside the cavity has an equilibrium energy which derives from the vector sum of all possible vibrations that can occur between the walls of the cavity. Each of these degrees of freedom contributes an amount $\varepsilon = kT$ to the total energy. To calculate the number of vibrational modes is a problem in pure geometry and the calculated Raleigh-Jeans curve is the correct prediction of the energy distribution in the hollow cavity according to the laws of classical physics. The prediction is correct at long wavelength (low frequency) and goes disastrously wrong at higher frequencies. It predicts an infinite contribution from high-frequency modes at any temperature. However, it is known from experience that most of the radiation at room temperature is in the infrared part of the spectrum and hence invisible. As the temperature is increased the oven starts to glow as described before, but higher-frequency contributions (ultraviolet and X-ray frequency) are only excited at much higher temperatures.

Simulation of the correct radiation curve was first achieved by Planck (1901) on empirical adjustment of the parameters in the Raleigh-Jeans formula. Instead of treating the vibrational modes as a function that varies continuously they were added as discrete elements of an infinite series. The crucial assumption was that rather than equipartitioned energies, each mode, treated as a discrete element, carried an amount of energy proportional to its frequency, i.e. $E = h\nu$. The proportionality constant, h, is known as Planck's constant. It follows immediately that at low temperatures there is insufficient energy to excite high-frequency modes. The revolutionary conclusion was that energy, like matter, consists of discrete elements, like atoms, called *quanta*. The energy of a monochromatic (single frequency) beam of radiation depends on the number of quanta, i.e. $E = nh\nu$, with n an integer. The quantum theory developed from this ground rule.

It is of interest to note that Max Planck probably never accepted the full consequences of his discovery. He certainly objected to the interpretation that electromagnetic radiation consisted of particle-like photons. It is ironic that this concept, proposed by Einstein to explain the photoelectric effect, returned to haunt him in later life when debating the interpretation of quantum theory with Max Born, Niels Bohr and others. Although it is generally agreed that Einstein lost the debate, he never conceded and history will probably prove him right. On another occasion when Hubble proposed an expanding universe Einstein had no hesitation in giving up his equilibrium model. In this action history will probably prove him wrong. His first instincts appeared to have served him well in both instances.

The appearance of integers in radiation events must have reminded contemporary scientists of the Balmer numbers. In fact, these numbers stimulated the Japanese physicist Hantaro Nagaoka to postulate (1904) that atoms are structured around a heavy nucleus, like the planet Neptune, with its rings simulating the lighter electrons orbiting the positively charged nucleus. It was only when a team of European scientists led by Ernst Rutherford and Niels Bohr made essentially the same postulate, ten years later, that the idea was taken seriously.

The concept is adequately illustrated by consideration of the simplest of all atoms, that of hydrogen. It consists of one massive proton with unit positive charge and one electron with unit negative charge.

Electronic transitions between the allowed orbits involve energies and radiation frequencies which could be shown in exact accord with Balmer's formula. The model shows why each element with its characteristic number of electrons has a characteristic emission line spectrum that matches a set of Fraunhofer lines in the solar spectrum if the element concerned is present in the solar corona.

Educated folk invariably have an idea of quantum mechanics and atomic structure, which more often than not implies some variation on Bohr's model of electrons orbiting an atomic nucleus. One of the many typical, but meaningless, models is shown in the diagram below.

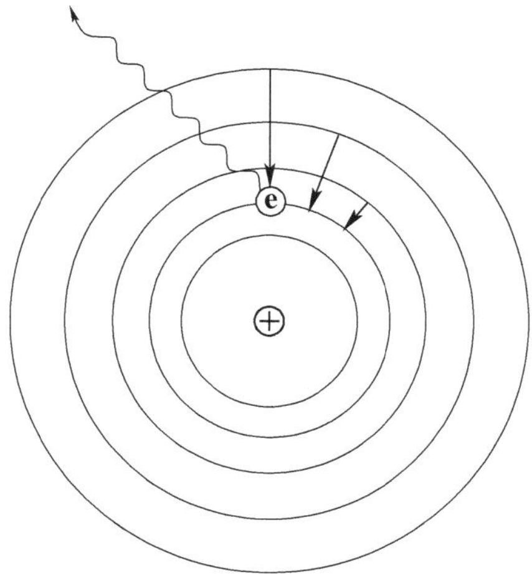

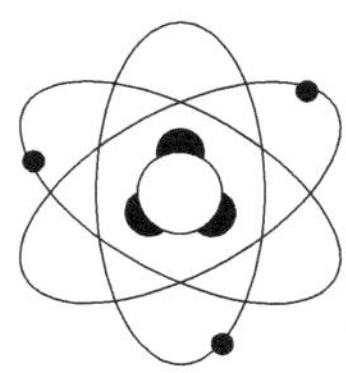

Although several flaws in Bohr's argument were pointed out soon after its first announcement the simple planetary analogy has now survived for a century and a year as the most popular of all atomic models. The fascination of the Bohr model lies therein that it elucidated three mysteries in one go: Rutherford's scattering experiments, The Rydberg-Balmer formulation of atomic spectra and the electrical properties of matter.

While the quantum theory was being formulated, theoretical advances with far-reaching consequences were made elsewhere. Advances which led to the formulation of the theory of relativity, the only theory that enquires into the nature of matter and its distribution through space.

The theory of relativity dictates that we live in a curved space of more than three dimensions. There is probably no concept in science that the layman finds more perplexing and, not surprisingly, constant efforts by anti-scientists to discredit the theory of relativity have continued for a century. The purpose of these efforts is not to gain an improved understanding of the theory or to propose an alternative interpretation thereof, both legitimate scientific pursuits, but rather to deny the complexity of the world by simply closing their eyes to reality. The reason why anti-science enjoys some respectability is because rigorous science often appears as counter intuitive to the uninitiated. The assumptions of Aristotelian physics demonstrate the situation well. What is more obvious and logical to the casual observer but, that heavier objects should fall down faster than lighter ones? For

Galileo to assert otherwise clearly identifies him as a rascal and an heretic. Aristotle's simple prescription was to elevate empirical awareness into the only validation of physics.

Anti-science at its most effective invokes religious dogma to stifle scientific enquiry. A familiar example is Aristotle's prime mover, the only source of mechanical interaction, associated with a perfect god. In this case the philosopher came to share in the divine authority, which for centuries defied the onslaught of experimental physics.

In the case of relativity and cosmology the two types of unreason are in cahoots to combat the spread of scientific understanding. On the basis of irrelevant empirical awareness, however naïve, relativistic phenomena are discussed pseudo-scientifically in non-relativistic context, for instance, to discount the perception of time dilation if, by definition, universal time flow is empirically known to be absolute and invariant.

Of more immediate concern is to show that without appreciating the importance of higher dimensions, called hyperspace, the relevance of relativity can never be understood. Not only does it escape the community known as the Flat-earth Society, but half of the world's population probably share their world view subconsciously. This means that they can never comprehend the most elementary notions of relativity theory and should be left in peace. With the more open-minded we now proceed to address the differences and theoretical possibilities of models in high-dimensional space.

Looking down from our lofty three dimensions we can imagine a pitiful two-dimensional world with all motion restricted to a flat surface. Immediately we understand how these individuals must have a mental block against picturing their surface as being curved. Only being aware of forward-backward and sideways motion, we can imagine their consternation on returning to their point of departure after a long arduous trek in a fixed direction. The more enlightened ones would conclude that their world was bounded and embedded in more than two dimensions, despite a pathological inability to envisage the third dimension. We have the same dilemma.

Having mastered the idea of existing on a curved two-dimensional surface, the next step is to contemplate the possibility that our three-dimensional vacuum may also be bounded and curved, rather than extending infinitely in three orthogonal directions.

Looking at the moon against a background of scattered clouds one gets the impression that the clouds are stationary and the moon is moving. By including some stationary object such as a tall tree in the field of vision it becomes obvious that the clouds are moving relative to both moon and tree. Looking from the window of a smoothly accelerating train compartment one gains the sensation of the entire station moving away from the train. These are two examples of relative motion, the description of which depends on the state of motion of the observer. Each observer considers its own frame of reference to be stationary and assigns any observed relative motion to the object under observation.

A traffic pointsman estimates the speed of two passing cyclists at 5 and 10 kph respectively. Each cyclist observes the other to approach at a speed of 15 kph. To avoid such confusion it is necessary to agree on a common frame of reference, which will be considered to be an inertial frame of reference. For the traffic on the street the appropriate frame of reference is clearly the solid earth. For interplanetary traffic it would be more appropriate to select the position of the sun as fixed and to measure all motion relative to that.

When electromagnetic theory predicts a constant speed of light, the question is, relative to what? Maxwell proposed that it was relative to the all-pervading ether,

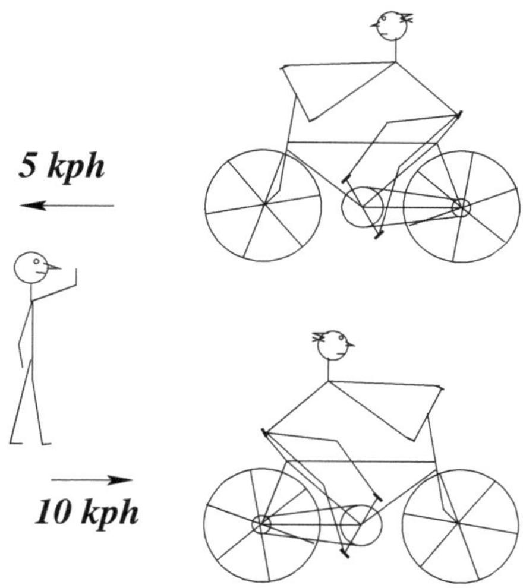

the medium in which the vibrations occur. This assumption creates the dilemma that different observers measure the same speed of a given beam of light, irrespective of their different speeds relative to the ether. The common-sense definition of relative motion, as worked out by the cyclists and the pointsman, can clearly not account for the speed of propagation of electromagnetic waves. In technical jargon, the equations worked out by Galileo to explain the different speeds on transformation between moving frames, are inadequate to deal with speeds approaching the speed of light.

To arrive at a mathematical formulation of relative motion that allows for a constant speed of light (c) it was necessary to involve a time parameter, together with the space parameter in the direction of motion. The result, known as the Lorentz transformation, has some revolutionary consequences. It shows that the length of a moving object contracts in the direction of motion, while its mass increases and clocks appear to be slowed down compared to stationary ones. To visualize these effects, imagine an intelligent creature that lives in a flat two-dimensional world, unaware of the concepts up and down, but fully aware of the invariance of a vector when rotated about point C. A three-dimensional intelligence interprets this as rotation about an axis through point C and perpendicular to the plane $X - Y$. To make mischief a three-dimensional rogue may tilt the rotation axis out of its perpendicular direction. On rotation, point A now moves out of the $X - Y$ plane and the two-dimensional being, who only observes the projection of point A in his planar

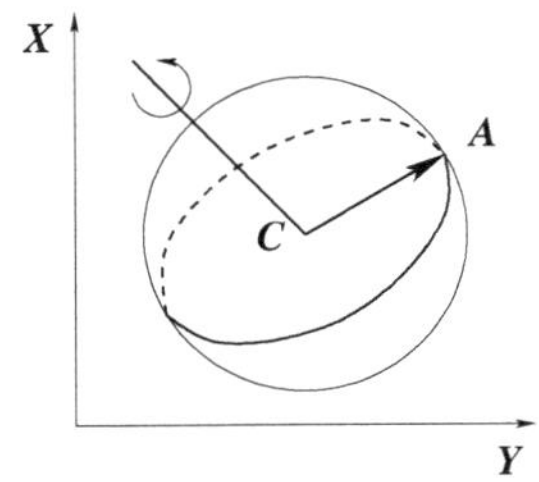

world, finds that it describes an ellipse, with the radius vector no longer invariant, as it dips below the plane or rises above the plane for half a cycle. Although he finds the observation hard to explain his only possible conclusion is that rotation of point A causes contraction of the vector.

Three-dimensional beings find themselves in the same dilemma when they observe relativistic rotations. Rather than the three-dimensional vector they observe a four-dimensional invariance, which now involves the time:

$$(\Delta r)^2 = (\Delta x)^2 + (\Delta y)^2 + (\Delta z)^2 - (c\Delta t)^2$$

Like their cousin in Flatland they notice a contraction of the three-dimensional vector, but compensated by time dilation, caused by rotation into a fourth dimension which can only be imagined.

The perceived contraction and dilation are not physically real processes, alternatively described as a change of scale [1].

The cosmological importance of Einstein's equation is that it identifies nuclear reactions as the source of stellar energy. The equally important conclusion of the four-dimensional structure of space-time is however, largely overlooked in standard expanding-universecosmology.

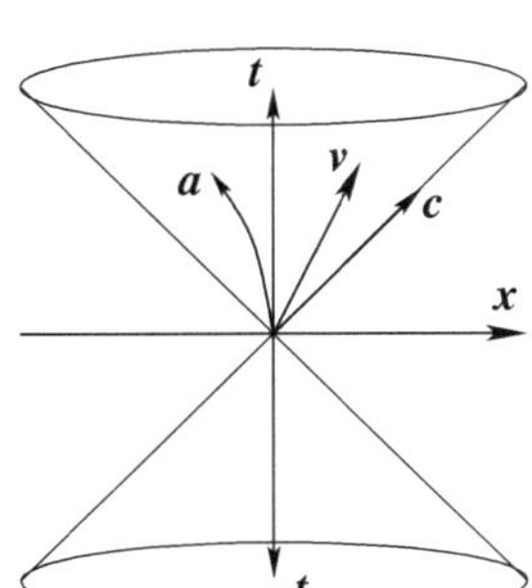

A Minkowski diagram is used to picture four-dimensional space-time. The common directions of three-dimensional space are represented together on a single axis, x, perpendicular to the axis that measures time flow. The Minkowski diagram, drawn in two dimensions is relatively easy to understand, but impossible to visualize. The problem lies therein that the three axes in space, normally denoted X, Y and Z are all directed at right angles (orthogonal) to each other. Although the expression:

$$\Delta r = \sqrt{(\Delta x)^2 + (\Delta y)^2 + (\Delta z)^2 + (\Delta w)^2} \tag{4.1}$$

gives a simple algebraic description of a line element in four dimensions it is geometrically impossible to construct in three. There is no room for the W-axis. In order to handle the concept of four-dimensional space it is therefore necessary to accept that it is humanly impossible to visualize more than three orthogonal directions only because our senses have evolved in local three-dimensional space, but our imagination is free to roam over any number of dimensions.

Any stationary object is said to follow a world line, parallel to the time axis. This means that the three space coordinates remain fixed as time changes. All motion is confined to occur within the time cone. An object that moves with uniform velocity v, has a world line at an angle with the time axis, but always in the time cone. Light beams have world lines in the surface of the light cone. Points in the light cone have specific values of all spatial and of the time coordinate, and describe events. Events in the past are mapped in the negative part of the time cone. In this representation the negative part of the time cone is as real as the positive part. Objects that move backward in time have world lines in the negative cone and waves moving back in time are mapped in the surface of the negative cone.

The theory of special relativity dictates that all physically real events occur in the time cone and are therefore called *time-like*. World lines of tachyons, or objects moving faster than light are outside the cone and called *space-like*. In a mathematical sense space-like and time-like events are equally real.

It is important to note that the restrictions of special relativity only refer to uniformly moving objects, which excludes accelerated and rotational motion. The world line (a) of an accelerated object appears curved in a Minkowski diagram and there is no obvious reason why such an object should not move out of the time cone.

The theory of general relativity was developed to deal with accelerated motion. By Newton's laws an accelerated mass experiences a force $F = ma$. Using a torsion balance the Hungarian physicist, Roland von Eötvös (1848–1919) could demonstrate that inertial mass is the same as gravitational mass. The general theory therefore became a gravitational theory as well. Einstein managed to prove by a simple argument that a gravitational field cannot exist in simple Euclidean space: Consider two coordinate systems, K and K', with a common origin—one of them stationary and the other rotating (accelerated) about the common Z-axis, in a space free of gravitational fields. A circle around the origin in the $X - Y$ plane of K is also a circle in the $X' - Y'$ plane of K'. Measurement of the circumference S and

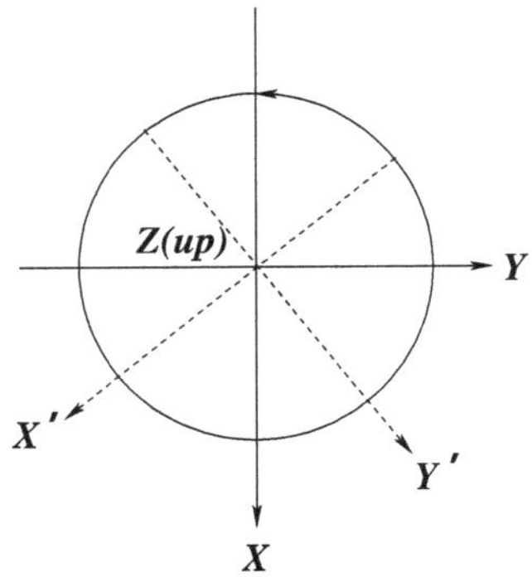

the diameter $2R$ in the stationary system must yield $S/2R = \pi$. However, when measured in the rotating system $S'/2R' > \pi$. The reason is that the measuring rod that moves with the rotating circumference appears to suffer Lorentz contraction, but not for measurement along the radius. The only interpretation is that the geometry of the accelerated system appears *non-Euclidean* and this fixes the geometry required to describe the gravitational field.

The geometry familiar to most readers was developed in classical Greece and codified by Euclid of Alexandria (*ca.* 300 BCE). It is based on a number of axioms of which the most famous is the 5th axiom of parallel lines:

> Given a line AB and a point P not on it, exactly one parallel can be drawn to AB through P.

It differed from the other axioms in not being *self-evident*. Generations of geometers engaged in a fruitless quest to either disprove the proposition or, alternatively, derive it from more primitive axioms. It is now appreciated that the concept of parallelism is closely associated with the fictitious idea of infinity. Accepting axiom 5 as valid leads to Euclidean geometry, characterized by the Pythagorian line element in n dimensions. Assuming that the pair of lines in axiom 5 may intersect at infinity, leads to a geometry with no parallels, known as elliptic geometry. The third possibility in which many different lines through P are parallel to AB leads to hyperbolic geometry. The latter two geometries are both known as non-Euclidean.

Elliptic geometry in two dimensions may be visualized on the surface of a sphere, such as the planet. The shortest distance between two points in such a surface is the segment of the great circle that connects the two points, and is known as a geodesic line. Any pair of great circles intersect somewhere in the surface and hence there are no parallels, except on a very small scale, such as the pseudo-Euclidean living space on the surface of the earth. The three angles of a triangle defined by geodesics now define the inequality $A + B + C > \pi$, in radian measure, as found by Einstein for relativistic rotation.

The important generalization of this concept from a two-dimensional curved surface to one of three dimensions was made by Bernhard Riemann (1826–66). The hypersurface on which four coordinates satisfy the condition $x_1^2 + x_2^2 + x_3^2 + x_4^2 = 1$ (compared to $x^2 + y^2 + z^2 = 1$ for the surface of a unit sphere) in Euclidean four-dimensional space, represents a non-Euclidean three-dimensional subspace. If this continuum represents the astronomical space in which man lives, the unboundedness of space does not necessarily imply infinitely long lines. A sufficiently powerful telescope could then enable an astronomer to observe the back of his own head. This idea, that space could be unbounded without being infinite was adopted by Einstein in his general theory of relativity.

If a system is accelerated by gravitation it means that the presence of gravitational mass distorts the geometry of space-time. Without going into mathematical details it should be obvious that the existence of matter causes space-time to curve. In fact, there is a reciprocal relationship between matter and curvature.

The gravitational equations cannot be solved without knowing the metric tensor. This is not a trivial problem, as shown by the statement [2]:

> "... we are still ignorant of the large-scale topology of the physical three-dimensional space and of the four-dimensional space-time manifold".

The very human response to forget about simplifying assumptions can, in time, dignify half-truths with undeserved authority. Most commentators with some knowledge of astronomy and cosmology feel convinced that the expanding universe is rigorously predicted by general relativity—it is not. Still, even without detailed solution the equations demonstrate some general principles of relativistic cosmology qualitatively correctly.

In a qualitative sense the effect of gravitational mass is to modify the local curvature of space and with it, the gravitational field, as shown in the schematic diagram.

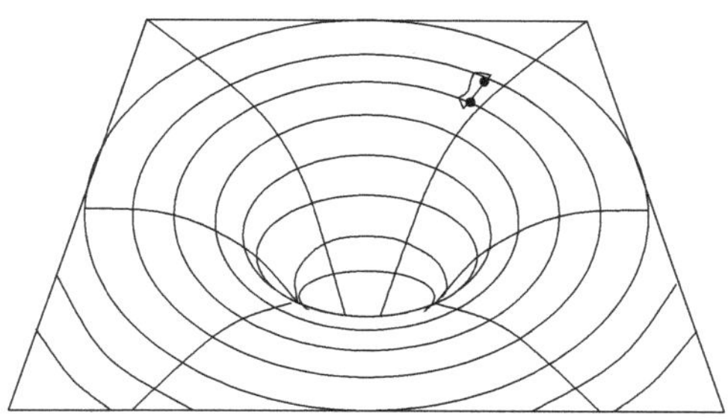

Any object that finds itself on the slopes of the gravitational field, that curves into the massive object at its centre, spontaneously moves along the changing gradient towards the attractor. This is not action at a distance, but apparent attraction, mediated by the curved space. Instead of moving uniformly along a straight line as in a Euclidean inertial space, it becomes accelerated along a curve. The exact curve is specified by the metric tensor of the non-Euclidean geometry, known as the geodesic. For the first time we get some understanding of Newton's first law and of why inertial motion should persist indefinitely. General relativity gives a simple answer: In a region of uniformly curved space a test object stays on the geodesic that follows the curvature. In the absence of gravitational effects, motion of the object, like the curvature, remains uniform and there is no distinction between the states of rest and motion, except in a gravitational field. In the vicinity of gravitational mass the uniform curvature is disturbed and accelerated motion occurs down the gravitational gradient.

Einstein's announcement of the theory of general relativity initiated a surge of activity aimed at finding a procedure to relate the electromagnetic field to the same space-time structure as gravity. Weyl's early attempt to derive the field from the gauge principle was only partially successful. When reformulated in terms of a complex phase factor the correct field could be predicted, restricted to wave motion, and without producing a unified field in one operation.

The first success was achieved independently by Kaluza and Klein by the addition of an extra space dimension that generated an additional five parameters from the field equations—to accommodate the electromagnetic field, with one to spare. The fifth dimension was postulated to be 'curled up' in cylindrical mode and hence unobservable.

Soon after the same result was obtained by Oswald Veblen, using the set of five homogeneous coordinates in four-dimensional projective space-time. Although, in this case there was no mysteriously curled-up dimension, the consensus was that it produced nothing beyond the mathematically simpler Kaluza-Klein model, and was not developed any further. By modelling the motion of a charged electron it was, however, shown [3] that the resulting equation could be reduced to the Klein-Gordon wave equation in tangent space, by simple rescaling. Although not appreciated at the time it has now been noticed [4] that the required scale factor is a simple function of the golden ratio.

The evidence becomes overwhelming that space-time is involuted in projective topology with general curvature related to the golden ratio. Not only does it account directly for the relationship between matter and antimatter, but also defines self-similarity, through the golden ratio, as a function of space-time curvature. Based on this, elementary number theory has been shown [5] to simulate the correct structure of the periodic table and a range of chemical properties, from atomic structure, to valence state, covalence, bond order, electronegativity and chemical reactivity.

More recently it was argued by Sachs [6] that Einstein himself, in later life, developed some unease over the exact formulation of his field equations, as evidenced by his comments [7]:

> Not for a moment, of course, did I doubt that this formulation was merely a makeshift in order to give the general principle of relativity a preliminary closed expression. For it was nothing more than a theory of the gravitational field, which was somewhat artificially isolated from a total field of as yet unknown structure.
> To remain with the narrower group and at the same time to have the relativity theory of gravitation based upon the more complicated tensor structure implies a naïve consequence.

It is then pointed out [6] that the symmetry group that underpins the field equations are too restrictive. The equations are covariant with respect to the continuous transformations in space-time, as they should be, but in addition, are also covariant with respect to discrete space-time reflections ($x^\mu = -x^\mu$), which is not required. This means that Einstein's tensor equations are not the most general expression of his theory.

Sachs [6] demonstrated that the irreducible representations of the unrestricted group obey the algebra of *quaternions*, behaving like second-order *spinors*. This quaternion metric field has 16 independent components compared to the traditional metric tensor with 10 components. The corresponding spin curvature tensor is antisymmetric as in Veblen's projective space-time. The 16 relations reflect the unified matter field theory anticipated by Einstein.

The theory of relativity remains controversial and to many minds plagued by paradoxes, such as the twin paradox and related versions thereof. Without going into details, all of these can be recognized as arising from a confusion between perception in tangent space and physical events that unfold in four-dimensional space-time. The simple example, given before, of the way in which the projection of a three-dimensional rotating vector appears to contract in flatland, directly explains all of these paradoxes.

References

1. M. Sachs, *Concepts of Modern Physics* (Imperial College Press, London, 2007)
2. R. Adler, M. Bazin, M. Schiffer, *Introduction to General Relativity* (McGraw-Hill, New York, 1965)
3. O. Veblen, B. Hoffmann, Projective relativity. Phys. Rev. **36**, 810–822 (1930)
4. J.C.A. Boeyens, *Chemical Cosmology*. Springer.com (2008)
5. J.C.A. Boeyens, P. Comba (eds.) Electronic structure and number theory. Struct. Bond. **148**, (2013)
6. M. Sachs, *General Relativity and Matter* (Reidel, Dordrecht, 1982)
7. A. Einstein, in *Albert Einstein Philosopher-Scientist*, ed. by P.A. Schlipp (Open Court, La Salle, Illinois, 1948)

Chapter 5
Delicious Mysteries

In order to get on with the job a new generation of postwar physical scientists had no choice but to accept the Copenhagen version of quantum mechanics, with all its warts and blemishes, as a working model. Not only was it necessary to convince themselves of the validity of their theory, but also to convince a sceptical world. They could claim the successes of wartime science to bolster their confidence and drive a public-relations campaign. They had a head start in the preposterous claim of Heisenberg and Born at the 1927 Solvay Conference and the apocryphal defeat of Einstein in debate with Bohr.

Quantum theory was riding high and the time was ripe to remove any lingering doubts. Under an americanized name the mathematician John von Neumann gained such renown as an advisor to the Manhattan Project that his analysis of the foundations of quantum mechanics became of seminal importance. In essence, this analysis provided mathematical 'proof' of Heisenberg's claim that "the fundamental physical and mathematical assumptions" [of quantum mechanics] "are no longer susceptible of any modification". The clarion call of "shut up and calculate" became the touchstone of modern science. Zero patience with 'futile' philosophical niggling over the absolute authority of the theory became the hallmark of this period. With physicists no longer engaged the rôle of developing a quantum world view was relegated to mystics and dilettanti with a poor understanding of science.

In academic circles it became necessary to dress up the wild assumptions of Copenhagen physics in a more palatable form, culminating in the idea of a non-Boolean quantum logic. This opened the floodgates towards development of a new tradition that explained any puzzling feature of the new paradigm as a 'quantum effect' that needs no further elucidation. The origins of quantum theory in two classical spectroscopic problems were conveniently forgotten and the new textbooks explained quantum mechanics as a complete refutation of classical physics.

In reality it had been known for many years that the light emitted by gases such as hydrogen, in an excited state at high temperature or electric discharge, rather than display a continuous spectrum, like a rainbow, it decomposes into widely separated discrete wavelengths. The demonstration by a high-school teacher, Johann Balmer,

J.C.A. Boeyens, *The Quantum Gamble*, DOI 10.1007/978-3-319-41621-2_5

that the observed pattern could be rationalized by a simple formula that contains a regular series of integers was not accepted by the experts as the basis of an explanatory physical model and discounted as an illogical use of numerology.

A more serious problem arose when recognized physicists failed to find an explanation of the observed spectrum radiated by a heated oven. Calculation of the radiation density according to the Boltzmann distribution of statistical thermodynamics, integrating over all allowed modes of vibration, resulted in the ridiculous prediction of infinite short-wavelength radiation density. The correct formula was obtained when Planck, in desperation, replaced the integration by summation of an infinite series over discrete frequencies.

In retrospect there is nothing mysterious about either observation. Both of them deal with electromagnetic waves and the prime characteristic of all waves is their periodicity, which is correctly described by integers. In the case of the Planck problem the integers are associated with the allowed wavelengths and frequencies of standing waves in a hollow cavity, or a solid emitter like a heated block of steel.

Following the pioneers of quantum mechanics, excluding Schrödinger and Planck, academics rejected the wave model and insisted on analyzing the observed phenomena in terms of particle mechanics, which inevitably leads to quantum probabilities, uncertainties, quantum jumps, confusion between subject and object, the mysterious role of an observer, half-dead cats, particles with wave properties, instantaneous non-local interaction, rejection of an underlying physical reality and causality—all of this without a shred of independent supporting evidence.

The preferred interpretation assumed that a quantum electron can be at more than one position at the same time, leading to the notion that a single electron is spread over the entire universe. To reconcile this proposition with the measurement problem it was necessary to assume that any quantum observation splits the universe into the system selected by the measurement and an infinitude of unobserved potentialities or parallel universes.[1] This so-called many-worlds interpretation has now developed into a serious quantum cosmology known as the multiverse, criss-crossed by a nest of wormholes.

It is by no means only physicists and cosmologists who subscribe to this moonstruck paradigm. The dilettanti and mystics of the world, not to mention chemists and psychologists, have developed these weird notions into a quantum philosophy in which anything goes. They live in a non-deterministic world without causality or reality, with total uncertainty and unpredictability. A world in which electrons have a free will, but in which the outcome of a quantum measurement depends as much on the system under investigation as the consciousness of the observer, without distinction between subject and object, or between cause and effect. The measurement creates a reality based on non-Boolean logic according to which a flipped coin can show heads and tails at the same time. An extensive

---

[1]It was at this point where a moon could be created or destroyed by a mouse winking its eye that Einstein finally abandoned the Copenhagen quantum model.

folklore has developed into a subculture based on the delicious mystery of half-baked quantum concepts.

The public perception that quantum mechanics portrays the workings of Nature as weirdly irrational was encouraged by the publication of a world bestseller designed to demonstrate that modern physics operates at the same level as oriental mystics [1]. The author was one of a community of hippies, qualified as physicists, operating from the Californian Bay area in the 1970s and 1980s. They used quantum theory to investigate phenomena such as the spoon-bending abilities of psychics, the physics of non-local interaction and the enhancement of research ability by LSD intake [2].

The Tao of Physics [1] is based on the myth that

> ... in atomic physics the truth is hidden in paradoxes that could not be resolved by logical reasoning.
> Like the mystics, physicists were now dealing with a nonsensory experience of reality and, like the mystics, they had to face the paradoxical aspects of this reality. From then on therefore, the models and images of modern physics became akin to those of Eastern philosophy.

At this point the author should have realized that the groping and sensational pronouncements of the Copenhagen pioneers do not necessarily equate to modern physics. If indeed physics is that closely akin to oriental mysticism it can no longer be science. The rationale was based on the statement:

> In modern physics, the universe is thus experienced as a dynamic, inseparable whole which always includes the observer in an essential way. In this experience, the traditional concepts of space and time, of isolated objects, and of cause and effect, lose their meaning. Such an experience, however, is very similar to that of the Eastern mystics.

Having stated without reference that

> ... we have to remember that particles are represented, in quantum theory, by wave packets ...

the author continues to explain quantum uncertainty in terms of a particle that 'moves around' at high speed within a wave packet—(probability wave packet?). Having discredited the Newtonian 'force of gravity, ... rigidly connected with the bodies it acted upon', as 'a strange hypothesis', he goes on:

> In field theory, the forces between particles appear as intrinsic properties of the particles

And in order to demonstrate consilience with *I Ching*:

> [The natural] laws are not forces external to things, but represent the harmony of movement immanent in them

whatever that means.

Details of the referred oriental philosophies are irrelevant in the present context. Of more importance is the caricature of modern physics, used in the comparison. Quantum theory is reduced to a paraphrase of some utterances by Heisenberg and Bohr in terms of probability waves, held mysteriously responsible for the behaviour of non-existent elementary particles, including electrons. The entire analysis hinges on inexplicable 'quantum' and 'relativistic effects', noting that

> Eastern mysticism has developed several different ways of dealing with the paradoxical aspects of reality

without any mention of Schrödinger's wave mechanics nor Lorentz transformation. The paradoxes exist largely in the author's mind.

At last count something like 43 editions of the book had been published in 23 languages. Enough to convince a sizable population of dilettanti that modern physics is best explored by meditation.

Physicists who find themselves scandalized by Capra's audacity have the trusty antitoxin of denouncing him as a crackpot, to rely on. The parallel with modern physics that he draws, is certainly naïve, but there is no denying the mystic elements in Copenhagen quantum mechanics. He could have made a more convincing case without involving the theory of relativity. It will be argued that the surprises of relativity are a valid feature of physics in four dimensions, rather than of wild assumptions of the Copenhagen type. Capra's interpretation may be suspect, but his data are sound: there is no rational understanding of quantum mechanics.

But quantum mechanics does not need Fritjof Capra to expose its fallacies. The theoretical chemists, Linus Pauling and C.A. Coulson performed adequately, without trying, from opposite sides of the Atlantic. The 'bunny ears' on atoms that interact according to their prescription are more toxic than eastern mysticism. Reluctant to abandon the convincing model for chemical interaction suggested by Sommerfeld's description of the electronic distribution in elliptic orbits around an atomic nucleus, these gentlemen devised an interpretation of Schrödinger's wave-mechanical analysis of the hydrogen atom to forcefully mimic the Sommerfeld results. It is necessary to repeat Pauling's statement that:

> ... if quantum theory had been developed by the chemist rather than the spectroscopist it is probable that the tetrahedral orbitals ... would play the fundamental role in the theory, in place of the s and p orbitals

Nobody argues with the most decorated scientist of the twentieth century and this crass oracular statement still inspires the chemical fraternity. The observation that urged this pronouncement is that, contrary to the four equivalent (degenerate) elliptic Sommerfeld orbits, directed towards the corners of a tetrahedron on the carbon atom, Schrödinger's equation predicts three degenerate (p) solutions and another (s) of higher symmetry at lower energy. In order to demonstrate that the Schrödinger solution does in fact 'confirm' the Sommerfeld picture it was necessary to ignore all rules in the book.

First, it was necessary to impose the required shape on the complex mathematical solutions of the wave equation. These *wave functions* have no direct geometrical meaning and in a sense correspond to the irrotational harmonic distortion modes in the surface of an incompressible fluid that symmetrically surrounds a central attractor. Such a construct closely resembles the hypothetical water planet that was analyzed by Laplace as a model to account for oceanic tides. All possible distortions are governed by spherical and circular modes, or linear combination of these.

What Pauling did, was to make linear combinations to generate linear distortions that could be rotated so as to constitute a set of orthogonal vectors to serve as a basis for the tetrahedral directions of Sommerfeld's elliptic orbits for carbon. He called these artificial constructs, *orbitals*, to emphasize their rôle as pseudo orbits.

But Pauling did not keep his eye on the ball. Each wave function has a characteristic set of *quantum numbers*, subject to a strict empirical rule, known as the exclusion principle. What he should have realized was that by rotating the polar direction of the wave function, he overstepped the exclusion rule. This is fatal. The famous operation of *orbital hybridization*, based on Pauling's rotation, therefore has no quantum-mechanical meaning. For more than fifty years the chemists of the world have been barking up the wrong tree.

The exclusion principle is as ruthless as the second law of thermodynamics. Without that there is no quantum mechanics. Not only undergraduate teaching, but all of the pretentious variants of computational quantum chemistry are based on the same linear combination of real one-dimensional functions. This is miles away from non-classical quantum theory. The crazy edifice known as either *ab initio* quantum chemistry or as density functional theory is based on sand and should be put to the sword.

With that out of the way chemistry can start with a clean slate in order to develop a theory of matter based on valid non-classical principles, rather than mysteries and magic.

A bewildering aspect of this situation is to find well-informed people refusing to take serious note of this argument against hybridization. They respond with rebuttals such as, "quantum numbers are not required in this instance", or "the exclusion principle should not be taken too literally". Even: "with proper understanding of quantum theory there is no problem".

In mitigation it must be pointed out that a set of orthogonal linear distortions, such as p_x, p_y, p_z, and their linear combinations are alternative solutions of Laplace's equation. Unfortunately that does not meet the requirements of a quantum theory with spin, and hence an exclusion principle. In a strictly classical model such as the Lewis cube, these orbitals could be of some use, provided the spin argument is not invoked. There is another proviso. The harmonic solutions of Laplace's equation are *surface* harmonics. They operate in the surface of a unit sphere and, unlike Pauling orbitals, do not extend to a nodal point at the origin. There goes their charm as bonding parameters.

A more recent source of quantum mystery is the celebrated demonstration by John Bell [3] that Einstein was wrong in challenging the non-local nature of quantum mechanics. Starting from the Copenhagen assumptions Bell managed to demonstrate by a convoluted mathematical discourse that the only way to account for the correlation of spin measurements on widely separated electrons was in terms of instantaneous non-local communication. It was triumphantly concluded that local realistic theories cannot account for such interaction. This is precisely what Einstein said without computational gymnastics.

Bell's contribution to the debate was to recommend experimental testing of the presumed correlation. He added the secondary unspoken condition that the quantum description of electron spin as a one-dimensional vector was unassailable. Experimental measurement of such a pair of spins must therefore constitute a conclusive verdict in favour of one or the other of the two theories: if a correlation is observed, quantum theory triumphs; if not, relativity prevails. This is impeccable reasoning provided the spin model is correct.

There is no fundamental guarantee that the *ad hoc* interpretation, offered by Pauli and Dirac to account for the empirically observed phenomenon of electron spin, is above reproach. By keeping an open mind on the issue, the many observations quoted by Bell [3], in support of Copenhagen theory, only confirm that the expected correlation exists. Without a full relativistic analysis of spin it is premature to discount Einstein's objections on the basis of an experimentally observed spin correlation.

The mystics and dilettanti did not have the patience to wait for a more detailed analysis of the issue. The so-called 'Bell's inequality' was universally accepted as absolute proof of non-local effects. One hesitates to allude to the *reality* of non-local effects, in view of the quantum-mechanical denial of reality. But this adds another dimension of importance to the non-local quantum mystery. Bell's [3, p. 155], ultimate conclusion concedes that

> ... finally, it may be that Bohr's intuition was right—in that there is no reality below some 'classical' 'macroscopic' level. Then fundamental physical theory would remain fundamentally vague until concepts like 'microscopic' could be made sharper than they are today.

On the topic of spin measurement Bell [3, p. 142], stated that:

> Phenomena of this kind made physicists despair of finding any consistent space-time picture of what goes on the atomic and subatomic scale. Making a virtue of necessity, and influenced by positivistic and instrumentalist philosophies, many came to hold not only that it is difficult to find a coherent picture but that it is wrong to look for one—if not actually immoral then certainly unprofessional. Going further, some asserted that atomic and subatomic particles do not *have* any definite properties in advance of observation. [...] Indeed even the particles are not really there.

Bell performs better in his analysis [3, p. 159], of von Neumann's contention that:

> ... the formalism of quantum mechanics is uniquely determined by these axioms; ... if a future theory should be deterministic, it ... must be essentially different.

Applied to spin eigenvalues Bell concludes, [3, p. 164], that von Neumann's

> ... 'very general and plausible' postulate is absurd.

So, maybe all of the delicious quantum mysteries are not as soundly based as generally presumed and that there could be some mileage in looking for an 'essentially different' theory.

The Quantum Demon is the subtitle of Kumar's historical account of quantum theory from Planck to Bell [4], without a word about quantum electrodynamics.

The sentiment is clearly in favour of Einstein, all the way up to the recognition of Bell's inequality. Reluctant to fall back on the Copenhagen orthodoxy, at this point, the author prefers Everett's multiverse to resolve the reality conundrum.

Everett's model, when stumbling over the measurement problem, but reluctant to underwrite the quantum gamble, interprets each possibility, predicted by quantum probability, to be realized in a separate universe. The result is an infinitude of parallel universes, multiplying all the time as new measurements are performed.

The way in which any given universe manages to stay on track over infinite-valued furcations to be navigated at each moment, is perhaps the deepest quantum mystery of them all—a real demon. Still, this is the point at which Kumar leaves his readership.

By no means is this the end of quantum mysteries but enough has been said to demonstrate that all of these can be traced back to a stubborn refusal to abandon the concept of point particles, or wave-particle duality, in favour of a wave model of matter. By following this alternative route to resolve the quantum mysteries, the argument leads into aspects of higher mathematics studiously avoided by many textbooks, but a summary of which appears unavoidable in order to explore the essential non-classical features of quantum theory.

References

1. F. Capra, *The Tao of Physics* (Shambhala, Boulder, 1975)
2. D. Kaiser, *How the Hippies Saved Physics* (W.W. Norton, 2011)
3. J.S. Bell, *Speakable and Unspeakable in Quantum Mechanics* (Cambridge University Press, Cambridge CB2 1RP, 1993)
4. M. Kumar, *Quantum: Einstein, Bohr and the Great Debate About the Nature of Reality* (Norton, New York, 2010)

Chapter 6
Non-classical Phenomena

Since the time of Newton the theories of physics have traditionally been defined in terms of higher mathematics. Paging through most treatises on modern physics the reader is confronted with page after page of mathematical symbolism that makes it virtually inaccessible to the uninitiated. It is left to a small number of dedicated theoreticians to scrutinize the validity of such mathematical material, after which it becomes widely, and usually non-critically, accepted and incorporated, in less rigorous form, into the textbooks of science.

The contributions of reputable scientists with a track record of original innovative work are, as a matter of course, more readily accepted than the ideas of an outsider. For the same reason the pioneering theories of celebrated scientists are seldom challenged. Should any defects become apparent, these are eagerly patched up by secondary assumptions. The rare instances in which fundamental theories are successfully challenged inevitably lead to major paradigm shifts. The two major theories of physics have, paradoxically, both survived all such attacks for a century.

The paradox involves the concept of instantaneous communication between quantum entities, which is rejected by the theory of relativity, but supported by quantum mechanics. Elementary logic demands that one, or both, of these need revision in order to eliminate this paradox. Fact is that the theories are destined to co-exist uneasily until the dilemma is resolved. The dilemma is exacerbated by the way in which both theories are ostensibly securely underpinned by impeccable mathematics. Mathematics of such complexity that, although well understood by professional mathematicians, it has to be accepted at face value by most physical scientists. On the face of it the two theories are formulated in terms of the same standard form of mathematics that prevents a comparison on this basis.

This problem is not as intractable as it seems and is readily resolved on noting that the disparate conclusions reached by quantum mechanics and the theory of relativity arise from two different, but legitimate, views of the physical world. Quantum theory followed the time-honoured approach of analyzing physical phenomena in three-dimensional Euclidean space, whereas the general theory of

J.C.A. Boeyens, *The Quantum Gamble*, DOI 10.1007/978-3-319-41621-2_6

relativity approaches the same situation as a problem in four-dimensional non-Euclidean space-time. The relationship between the two approaches reduces to local and global views of the cosmos. Topologists distinguish between tangent three-dimensional space and an underlying four-dimensional space-time.

It will be argued that this difference demands the use of fundamentally different mathematical analyses, based on traditional complex algebra and the hypercomplex algebra of *quaternions*, respectively. The different results that derive from these different algebras first became apparent on comparison of the mechanical effects predicted by the special theory of relativity and classical Newtonian mechanics. The effects, consistent with the Lorentz transformation of the special theory differ so radically from Newton's results that they became known as *non-classical* effects. A so-called classical *limit* is approached as the speed of an object of interest approaches that of a light beam, i.e. $v \to c$.

The pioneers of quantum mechanics postulated a comparable classical limit for a system in which Planck's constant tends to zero, i.e. $h \to 0$. This is clearly a meaningless proposition and a logically valid specification of a quantum-classical limit has never been found. The simple reason is that such a limit is undefined, because, like classical mechanics, traditional quantum mechanics is formulated in terms of complex algebra in three-dimensional Euclidean space. The only non-classical concept traditionally associated with quantum mechanics is electron spin. However, the spin variable does not emerge naturally, but has to be added empirically to 'quantum systems'.

It will be shown that spin occurs as non-classical action; a holistic mode of four-dimensional harmonics, known as a *spinor*, which also underpins the Lorentz transformation of special relativity. On treating both theories in four-dimensional formalism and without separation of space and time variables, they become fully consistent and the vexing non-local quantum effects that alienated Einstein, and the wave-particle duality that annoyed Schrödinger, disappear automatically.

The reconciliation between the theories of relativity and the quantum, in the form of four-dimensional holistic matter-wave mechanics, and the way in which this approach redefines non-classical phenomena is discussed in more detail below. The complication arising from the non-linearity introduced by the curved space-time of general relativity and which invalidates the superposition principle of pioneer wave mechanics, is discussed in the next chapter.

The Mathematical Model

The two theories differ in structure from classical physics in the essential use of *complex algebra* in their formulation. This esoteric form of algebra was discovered by mathematicians in their search for the square root of negative numbers. By recognizing the fact that unity has two possible square roots, because $1 \times 1 = -1 \times -1$, negative unity only has an *imaginary* square root, denoted by i. Since there is no room for i on the real line (R) of integers and fractions, including *irrational* fractions, mathematicians invented a two-dimensional number system in order to accommodate imaginary numbers.

The following (Argand) diagram shows how any real number x is converted into $-x$ by a rotation of 180° (π radians) about a perpendicular axis (Z) through the origin, and hence that x is converted into its imaginary[1] counterpart of ix by a rotation of $\pi/2$. This interpretation leads to the general rule of representing any rotation (θ) in terms of a *vector*, $z = a + ib$.

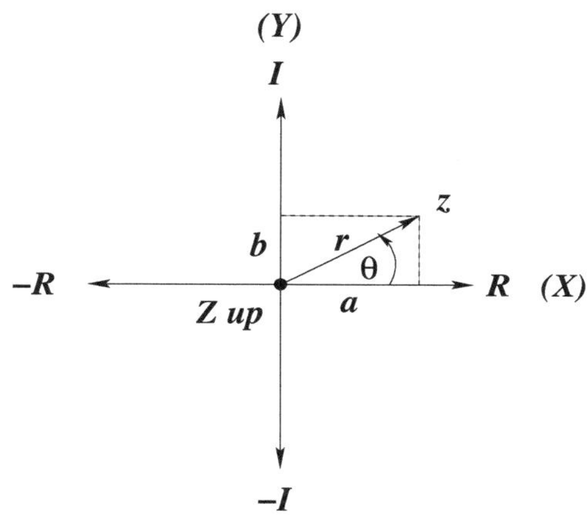

By elementary trigonometry the complex number z can also be formulated as

$$z = r(\cos\theta + i\sin\theta),$$

or in terms of Euler's formula, as an exponential function

$$z = r\exp(i\theta) = re^{i\theta},$$

where the irrational constant $e = 2.7182813\ldots$, is better known as the basis of natural logarithms. Pauling's rotation, mentioned in the previous chapter, involves multiplication by i, which transforms the quantum number m_l of his p_y orbital from 1 to zero, the same as for p_x and p_z; forbidden by the exclusion principle.

Using the same rules that apply in the one-dimensional algebra of real numbers, a two-dimensional algebra of *complex* numbers has been developed on the basis of the Argand diagram. A special property of this two-dimensional complex field is that it is closed under a set of *orthogonal rotations*, through π radians, about the axes X, Y and Z. A rotation about X that transforms I into $-I$, followed by a rotation

[1] I.e. two-dimensional.

about Y, which turns R into $-R$, is equivalent to a single rotation about Z. Such a set of operations, with *closure*, constitutes what is known as a mathematical *group*; a *rotational group* in this case.

In line with its *imaginary* connotation complex algebra remained an esoteric curiosity without any application in the *real* world of science. That is, until the discovery of relativistic and quantum effects early in the previous century.

The first to achieve mathematical formulation was the theory of special relativity, developed in order to reconcile the observation of a constant speed of light in the vacuum with the classical theory of relative motion. It came as a direct result of Maxwell's formulation of the electromagnetic field and mathematical formulation of the effect which became known as the Lorentz transformation. The remarkable feature of this theory was the necessity to mix space and time coordinates on an equivalent basis. In classical mechanics any motion can be described in terms of space coordinates that change as a function of universal time. This is not allowed by the Lorentz transformation.

The implication of the new theory was that relative motion entails, not only a change of space coordinates, but also of a time coordinate. The mathematical formulation of this condition is only possible in a four-dimensional space, with time as one of the variables, in the form of an imaginary quantity, *it*. All of the counterintuitive predictions of relativity theory can be reduced to this requirement.

A second surprise was soon to follow on discovery of the mathematical basis of quantum theory. This was foreshadowed by the first, unsuccessful, attempt by Herman Weyl to unify the gravitational and electromagnetic fields by a *gauge* theory. The world geometry behind the proposal allowed for a modification of the magnitude of a vector on displacement. It was left to the genius of Schrödinger to show how to rescue Weyl's theory by substituting a *phase* change for the proposed gauge effect. Unfortunately the original terminology persisted and the seminal idea is still referred to as a gauge principle. The phase change as envisaged by Schrödinger is commonly associated with wave motion, for which the observed intensity corresponds to the absolute square of the amplitude, on which the phase change has no measurable effect.

Be that as it may, whereas gauge variation is described by a real variable, the more relevant phase change is described by a complex function of the form $e^{i\alpha}$. The product $e^{i\alpha} \cdot e^{-i\alpha} = e^0 = 1$. In this form it led to the formulation of *wave mechanics* to account for quantum phenomena. Applied to an electron with a conserved charge it confirms that the electron cannot be a mechanical particle.

All attempts to formulate a quantum theory in terms of real variables have been unsuccessful and all efforts to transform complex quantum variables into real functions have produced distorted theories, such as the concept of wave-particle duality and orbital hybridization. As in the case of relativity theory quantum mechanics is characterized by many other unusual, counterintuitive features, invariable related to the complex formulation.

At first blush the theory of relativity and quantum mechanics appear to be based on the same complex algebra, without any hint of a discrepancy. However, the relativistic four-dimensional symmetry between space and time variables, not recognized in quantum theory, could account for the discrepancy. In this context it is significant to note that this same symmetry appeared in Maxwell's original formulation of the electromagnetic field, which resurfaced in Minkowski's interpretation of the Lorentz transformation as a four-dimensional complex rotation.

On recognizing the electromagnetic field as a four-dimensional construct Maxwell formulated the field mathematically in terms of quaternions. Posthumous simplication of Maxwell's formulation by Heaviside and Gibbs used scalarless quaternions, in the form of three-dimensional vectors. This formulation has been perpetuated in all textbooks, while the more detailed quaternionic field has been forgotten. This is the primary reason why relativity and quantum theories appear to be incompatible today.

To better appreciate the algebra behind the work of Minkowski and of Maxwell it is necessary to turn to Hamilton and his efforts to formulate rotations in three-dimensional space. His work consisted of a search for the three-dimensional equivalent of the two-dimensional group that describes the complex plane. Instead of a square he considered the rotation of a cube with coloured faces. His best result

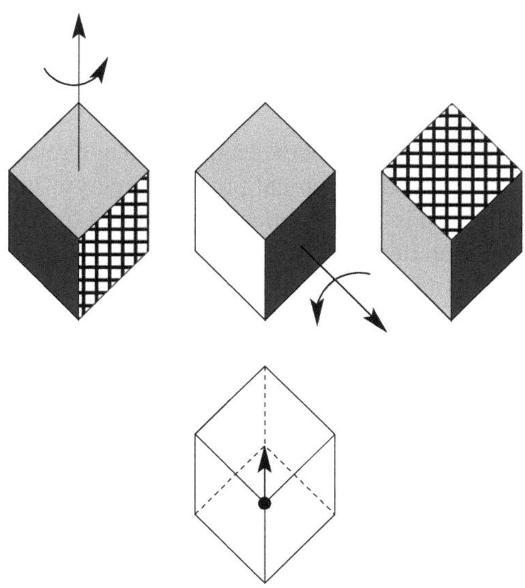

was to show how successive rotations of $\pi/2$ gave rise to an orientation which could be reproduced by a single rotation of $2\pi/3 = 120°$ about the body-diagonal through the marked corner, but not by a mathematical *group* of $\pi/2$ rotations.

In order to recover the group structure, Hamilton defined rotations in four-dimensional space and managed to reproduce the experimental result, observed in the rotation of the coloured cube, theoretically. A four-dimensional *hypercomplex* algebra that corresponds to this space has elements of the form

$$q = \alpha + i\beta + j\gamma + k\delta \equiv a_0 + \sum a_i e_i ,$$

known as *quaternions*. All of the a_i, $i = 1 \rightarrow 3$, are real numbers and the e_i are generalizations of $\sqrt{-1}$ in the form of square matrices.

In addition to some important engineering applications[2] quaternions are routinely applied in the physical sciences [1], mostly without acknowledgement, to problems in crystallography, the kinematics of rigid bodies, the Thomas precession, the special theory of relativity, classical electrodynamics, the equation of motion of the general theory of relativity and Dirac's wave equation. Most compelling in the present context is that the 4π intrinsic spin of quantum mechanics can only be described by a quaternion spinor.

On closer examination quaternions are seen to result from the combination of two complex numbers, in the same way that a complex number is constructed by the combination of two real numbers. This process, known as dimensional doubling, explains why the three-dimensional hypercomplex number, sought by Hamilton, cannot be defined.[3] The next higher hypercomplex number, known as an *octonion* consists of a single scalar (a_0) and a seven-dimensional vector.

The Physical Model

The physics behind the discrepancy between quantum and relativity theories is best illustrated in terms of a four-dimensional potential-energy function. The physical condition of an equilibrium is mathematically expressed by a differential equation in which the sum of second derivatives of the potential energy of a test particle in a field, with respect to all variable coordinates, vanishes to zero. In four dimensions this condition is expressed symbolically in the form:

$$\Box^2 V = \frac{\partial^2 V}{\partial x_0^2} + \frac{\partial^2 V}{\partial x_1^2} + \frac{\partial^2 V}{\partial x_2^2} + \frac{\partial^2 V}{\partial x_3^2} = 0$$

To understand the meaning of this hairy equation it is noted that the potential energy of a test particle in a field, such as the gravitational field, depends on the position of the particle and is modified by an applied mechanical force, F, which determines the *rate* at which the potential energy changes as a function of position. In differential notation this is described by the *first derivative*, $\partial V/\partial x = -F$. The rate at which the

[2]Outlined in Appendix A.

[3]The present author spent many fruitless hours on the same quest, before he became aware of Hamilton's result.

force changes, *i.e.*,

$$\frac{\partial F}{\partial x} = \frac{\partial}{\partial x}\left(-\frac{\partial V}{\partial x}\right) = -\frac{\partial^2 V}{\partial x^2}$$

and the condition of balanced forces now leads back to the equation over all coordinates.

Both theories, of relativity and the quantum, depend on modified forms of this fundamental harmonic equation. Where they differ is when it comes to solution of the equation. Solution of the relativistic equation leads directly to the Lorentz transformation, in the form of a *quaternion*. The scalar term, also known as the *versor*, represents time, which is inextricably associated with the space vector, and together they define the space-time rotation known as a *spinor*.

Quantum mechanics follows the Heaviside route in which the time parameter is separated from the space part. The time derivative

$$\frac{\partial^2 V}{\partial x_0^2} = \frac{\partial^2 V}{\partial (ict)^2} = -\frac{1}{c^2}\frac{\partial^2 V}{\partial t^2},$$

is no longer entangled with the space derivatives and the fundamental equation becomes a three-dimensional wave equation with the Laplacian,

$$\nabla^2 V = \frac{1}{c^2}\frac{\partial^2 V}{\partial t^2}.$$

Solutions to this equation, obtained by separation of the variables as shown in Appendix B, are no longer spinors, but vectors that rotate in three-dimensional space. Schrödinger's equation, which is fundamental to quantum mechanics, is of this type. When used to describe the hydrogen electron it defines a harmonic function, readily interpreted in terms of orbital angular momentum. To compensate for an unexplained, spectroscopically observed two-fold variability, called *electron spin*, it is empirically added on as a linear vibration, in an unspecified direction. In this form it cannot account for the four-dimensional *action*, defined by the quaternionic spinor.

When trying to account for spin correlation, observed on widely separated electrons, the assumed model of independent spins relies on *non-local*, instantaneous communication, mediated by wave-function collapse; at variance with the theory of relativity. This complication does not arise if the separated electrons are characterized by symmetry-related spinors. This symmetry remains conserved for previously spin-paired electrons, unless modified in an external field. There is nothing to explain, neither in terms of wave-function collapse, nor in terms of any other artificial construction.

The discrepancy between the two theories can now confidently be traced back to disruption of the quaternionic spinor on turning the space-time potential into a wave equation, with disastrous consequences. One instance where the three-dimensional approximation leads to a misleading interpretation has been identified, but there are many more. Many of the mysterious conclusions ascribed to 'quantum effects' could be of similar nature and deserve further scrutiny.

The Chemical Model

At the undergraduate level chemistry is taught as if underpinned by appropriate solutions of Schrödinger's wave equation. This is a fairy tale. In this context the three-dimensional harmonics are referred to as non-classical. However, the only non-classical attribute of an electron is the spin, which is only known empirically and has never been derived from first principles, not even by Dirac. All that remains in three dimensions is interpreted as orbital angular momentum, which implies rotation, in this case of a charged particle known as an electron. This is exactly the model proposed by Bohr to describe a hydrogen atom, and the reason why his proposal was finally rejected: A charged particle accelerated in order to maintain a stable orbit must radiate its energy away and spiral into the nucleus. It is never explained why the Copenhagen electron does not have the same problem.

In the same way that the complete separation of space variables results in a simplified solution that consists of three *real* functions, or orbitals, the three-dimensional Schrödinger harmonics are a simplification of a four-dimensional spinor. To be consistent with the observed periodicity of chemical elements the Schrödinger solution was upgraded by Pauli with the addition of a linear spin vector, directed with equal probability also in the opposite direction. Dirac managed to incorporate the spin vector more intimately with Schrödinger's angular momentum, that consists of a complex pair and a real polar vector.[4]

Some analysts have shown that by linking the polar and spin vectors into another complex pair (known as spin-orbit coupling) the total variability resembles a Clifford algebra (a scalarless quaternion plus independently directed universal time). Although the resulting spin function lacks the holistic generality of a hypercomplex spinor, it has the required structure to account for the spin correlation normally misinterpreted in terms of wave-function collapse, but this has gone unnoticed before.

The stepwise reduction of a spinor, from the hypercomplex to the classical, by progressive separation of the variables, is formally summarized in the following table:

[4]This was also achieved by Lévy-Leblond [2] in a non-relativistic analysis.

holistic spinor	$\boldsymbol{q} = \alpha + i\beta + j\gamma + k\delta$	Maxwell
spin+orbit	$\boldsymbol{L} = \alpha + i\beta\,,\ \boldsymbol{S} = \gamma + i\delta$	Dirac, Levy-Leblond
ang. mom., spin	$\boldsymbol{L} = \alpha + i\beta\,,\ L_z = \gamma\,,\ S = \delta$	Pauli
ang. momentum	$\boldsymbol{L} = \alpha + i\beta\,,\ L_z = \gamma\,,\ t = \delta$	Schrödinger
Kepler rotation	$\boldsymbol{R} = \boldsymbol{\alpha} + \boldsymbol{\beta} + \boldsymbol{\gamma}, S = \delta$	Sommerfeld
Space, time	$\boldsymbol{X} = \boldsymbol{\alpha} + \boldsymbol{\beta} + \boldsymbol{\gamma},\ t = \delta$	Pauling

With mathematics out of the way, the description of electron spin as a mechanical process can be considered in more detail.

Electron Spin

The concept of electron spin developed from the observation that in order to account for the periodic table of the elements two electrons had to be at each of the energy levels specified by the quantum numbers of Sommerfeld's atomic model. The multiplicity of a degenerate energy level of quantum number s was known to be given by $2s + 1$, which in this case implies $s = 1/2$. The two-fold variability observed in a directed magnetic field (Stern-Gerlach experiment) is then correctly specified by a magnetic quantum number $s = \pm 1/2$, and this *action* was later explained as *electron spin*. It was inferred that spin differs from rotation in a fundamental way.

There are two fundamentally different types of rotation *viz.* rotation about an axis and about a point; the latter known as *spherical rotation*, visualized by rolling a coin around a second stationary coin. The original arrangement is repeated only after rotation of 4π. It represents an element of four-dimensional symmetry, which

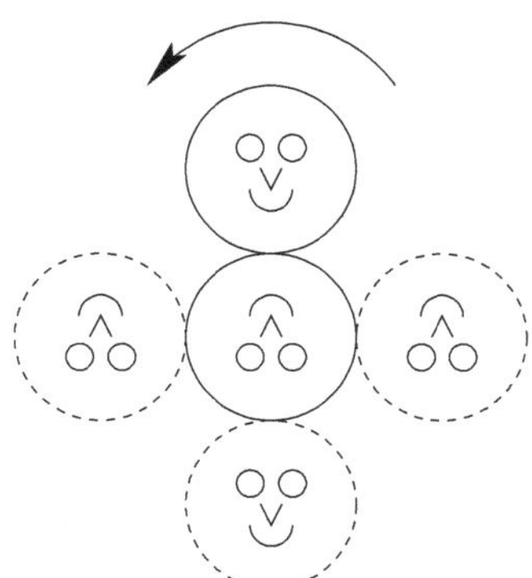

does not appear in three dimensions. The action known in quantum theory as *spin* is described by spherical rotation [3] and the rotational groups that describe the double cover are also known as spin groups. It is noted that the spin groups are only generated in normed division algebras in 4 and 8 dimensions, which explains why spin variables must be added empirically to 3D quantum models, such as Dirac's equation.

To solve Dirac's equation it is necessary to separate space and time variables, which destroys the quaternionic spin function, to be re-inserted by hand. The common textbook statement that spin is a relativistic effect results from a misreading of Dirac theory. The more correct statement is that spin is a four-dimensional phenomenon. It is measured in units of action, the same as angular momentum, which is two-dimensional.

The significance of division algebras as the basis of physical theory depends on the way in which they describe dynamic action in 2^n dimensions, $n = 0 \rightarrow 3$, as vibration, rotation, spin and *spin involution* [4].

Although quaternions represent four-dimensional vector fields, by projection into lower dimensions, they also serve to describe rotations in two and three dimensions as defined by the *special orthogonal groups*, SO(n). In technical jargon, each SO(n) group is 'doubly covered' by a special *unitary* group SU($n - 1$), called a spin(n) group of spinors. The Pauli spin matrices are representations of spin(3), which is a formal description of an empirically observed effect, called spin, not derivable from the 3D vector field. This is the form in which all chemists use and understand electron spin [5].

Although adequate for the discussion of chemical effects by experimentalists it is important to realize that **spin action** is a strictly four-dimensional phenomenon, the 3D projection of which only has qualitative validity. This fact is recognized by many theoretical physicists, but poorly appreciated by chemists. Most at risk are computational chemists who use linear three-dimensional differential calculus in chemical modelling. Spin(4), as a double cover of SO(4), will be shown to imply nonlinear projective topology that does not allow linear combination of atomic orbitals or wave functions.

The argument whether to use quaternions or vectors that raged in the late nineteenth century [6, 7] was settled in favour of the latter, as the mathematical basis of electrodynamics, rather than a *spinor* (quaternion) formulation of the Maxwell field. This debate is being reopened on a wide front [8] and chemical theory should take note of these trends.

The realization that complex algebra was of more than mathematical interest was treated with suspicion for a long time, and in many quarters still is. A specific instance of physical importance in which complex algebra is indispensible, is in fact quantum theory, as evidenced by the irreducible Hamiltonian equation of motion:

$$\frac{dx}{dt} = i\hbar(Hx - xH).$$

In order to identify physical systems that rely on hypercomplex algebra in the same way, it is only necessary to look at the theory of special relativity and spin.

A four-dimensional potential is represented in entangled Minkowski space-time by an equation with quaternionic solutions, which represent hypercomplex rotation. The importance of this result is that the Lorentz transformation is identically described by such a rotation. Even more important in the present context is that the four-dimensional action in this potential field is described by the spin function, identified before.

Two perennial mysteries of quantum theory are immediately resolved by this observation: the inability to derive electron spin analytically in three-dimensional wave mechanics and the unexpected commutation properties of quantum-mechanical pairs of conjugate variables. Quaternion variables do not commute. The obvious answer is to find quaternionic solutions of d'Alembert's 'matter-wave equation', also known as the Klein-Gordon equation, to replace traditional quantum theory. The theoretical justification of this proposal is that only the theory of relativity can account for the nature of the electromagnetic field and only in Minkowski space-time. All known quantum phenomena are intimately related to this same field and hence cannot be characterized correctly in three dimensions. The biggest potential bonus is that by identifying the common ground between the two theories the notorious incompatibility related to the so-called 'non-local' nature of quantum theory is also eliminated.

The standard non-local argument holds only if the orthogonal components of spin are independent and directed probabilistically. However, four-dimensional spin action is a conserved constant of the motion. A spinor, defined in quaternion algebra, does not depend on any measurement and is a holistic unit with fixed space-time chirality. The four-dimensional spin function is said to have chiral symmetry, like a pair of hands in three-dimensional space. Right and left hands are rarely confused, except in pharaonic Egypt.

A right hand, examined from any perspective, always exhibits its characteristic handedness, distinct from that of its mirror image. The pair of previously, chemically correlated spins on the separating fragments of the EPR thought experiment, likewise remain distinguishable by virtue of their opposite chiralities. A magnetic field, applied in a spin measurement, cannot change the chirality of the spin—it only identifies the component of spin in the field direction, but that is fixed by chirality. Whatever one observer finds on measuring the first spin, an opposite outcome is inevitably fixed elsewhere by the inverse chirality of the second electron. The chirality of both spins remains fixed without further assumption and the mystery disappears. Whatever is measured on one side is fixed to be observed as inverted on the other.

Measurement of electron spin is effected in an applied magnetic field that aligns one component of the spinor with, or against, the field direction. The alignment depends on the orientation of the spinor relative to the field, with 'up' or 'down' as the only possible outcome. The second spinor, with inverted chirality, must invariably yield the opposite result. The magnetic field does not create the spin and there is no communication between the two measurements. The experiment has nothing to do with the collapse of a wave function and the notorious EPR paradox has no meaning in four-dimensional space-time.

Furthermore, independent measurements of the two electron spins must always reflect the difference in chirality, irrespective of the relative orientation of the two applied magnetic fields. This is the situation shown by Bell [9] to be unreachable in a local realistic theory with probabilistic spin orientations. And this is exactly where he went astray:—the four-dimensional components of spins are not subject to probabilistic laws and Bell's inequality is irrelevant in the analysis of the EPR 'paradox'.

The way in which to exercise the quantum demon of the previous chapter has now been opened up. The issue is logically decided directly by the recognition of a four-dimensional reality. The concept of 'point particle' now acquires an additional dimension of zero time that denies the possibility of its existence. The joke is that the non-classical heir of quantum theory now favours space-time continuity.

Conclusion

In summary, the casual observer perceives a three-dimensional world and formulates a quantum theory of point particles, whereas the theory of general relativity stipulates periodic action in four dimensions. All that remains for a new quantum theory is to establish the relationship between action, matter and energy as a function of space-time topology.

It is noted that the point particle was smuggled into linear wave mechanics to exploit the superposition principle. It became acceptable to simulate wave motion by a probabilistic distribution of point particles that anticipates the outcome of any conceivable measurement, by creating it. Not only does it fail to capture the essence of waving, but it introduced the arcane notion of interaction between a conscious mind and probability waves, in order to explain an uncertain quantum reality. The obvious conclusion that a wave function describes a wave, disposes of this occultism and reveals the true nature of an electron.

Without probability waves there is no measurement problem nor quantum uncertainty. The reciprocity between conjugate variables is an inherent feature of any wave formalism [10]. There is nothing left to explain or get excited about because a wave is not confined to a point, but occurs in a finite region of space-time. Any measurement can now be performed without the risk of creating an entirely new universe. The measurement simply reveals what exists out there. It may even disturb the system of interest, but not to the extent of creating it from a probability wave.

Firmly convinced aficionados invoke [11] what they call 'quantum interference' experiments with a Mach-Zehnder interferometer to demonstrate the reality of wave-particle duality and the existence of parallel universes. However, it is readily shown [12] by considering reflection phase shifts that elementary optics demonstrates such claims to be 'grossly misleading'.

References

1. P.R. Girard, The quaternion group and modern physics. Eur. J. Phys. **5**, 25–32 (1984)
2. J.-M. Lévy-Leblond, Nonrelativistic particles and wave equations. Commun. Math. Phys. **6**, 286–311 (1967)
3. E.P. Batty-Pratt, T.J. Racey, Geometric model for fundamental particles. Int. J. Theor. Phys. **19**, 437–475 (1980)
4. J.H. Conway, D.A. Smith, *On Quaternions and Octonions* (A K Peters, Natic MA, 2003)
5. J.C.A. Boeyens, Understanding electron spin. J. Chem. Ed. **72**, 412–415 (1995)
6. M. Sachs, *Relativistic Implications in Electromagnetic Field Theory*, in [7], pp. 541–559
7. T.W. Barrett, D.M. Grimes (eds.), *Advanced Electromagnetism* (World Scientific, Singapore, 1995)
8. M. Sachs, *Quantum Mechanics from General Relativity* (Reidel, 1986)
9. J.S. Bell, On the Einstein–Podolsky–Rosen paradox. Physics **1**, 195–200 (1964)
10. J.C.A. Boeyens, *The Theories of Chemistry* (Elsevier, Amsterdam, 2003)
11. D. Deutsch, *The Beginning of Infinity* (Viking, New York, 2011)
12. K.P. Zetie, S.F. Adams, R.M. Tocknell, How does a Mach-Zehnder interferometer work? Phys. Educ. **35**, 46–48 (2000)

Chapter 7
Non-linear Phenomena

Searching for a theory of everything John Barrow [1] faces the age-old choice between atomism and holism. He opts for the former, but to avoid singularities and infinities he invokes *string theory* to define an atom as a one-dimensional object that wiggles in a void of many dimensions. He describes the continuous holistic alternative as:

> ... Nature operating holistically to produce a harmonious equilibrium in which every ingredient interacts with its fellow to produce a whole that is more than the sum of its parts.
> ... The holistic view assumes nature to be intrinsically non-linear so that non-local influences predominate and interact with one another to form a complicated whole.

Stating that:

> Only recently ... have scientists come to terms with the description of intrinsically complex non-linear systems.
> A successful study of natural laws needs to start with the simple linear problems if it is ever to graduate successfully to holistic complexities created by non-linearity.
> Linear problems are easy ... problems where the sum or difference of any two particular solutions is also a solution ...

He explains that:

> ... if a situation is linear or dominated by influences that are linear, it will be possible to piece together a picture of its whole behaviour by examining it in small pieces. The whole will be composed of the sum of the parts. ... Linear phenomena are thus amenable to very accurate mathematical modelling.
> Non-linear problems are none of these things. They amplify errors so rapidly that an infitesimal uncertainty ... can render prediction of its state worthless after a very short period of time.
> ... the simplest imaginable non-linear equations exhibit behaviour of unsuspected depth and subtlety which is, for all practical purposes, completely unpredictable.

He reaches the strange conclusion that:

> Despite the ubiquity of non-linearity and complexity, the fundamental laws of Nature often give rise to phenomena that are linear.

J.C.A. Boeyens, *The Quantum Gamble*, DOI 10.1007/978-3-319-41621-2_7

> It [linearity] is crucial in rendering the world intelligible to us . . .

He seems to be missing the point that his laws of Nature may merely be the laws of Physics which could simply be plain wrong and meaningless.

Non-Linear Quantum Chemistry

The development of quantum mechanics from a linear differential equation is seen as its major defect at the basis of physical science. Its most spectacular failure occurs in quantum chemistry that relies on the linear superposition of basis functions to simulate the formation of molecules by the nonlinear interaction between standing electron waves.

To address this problem it is necessary to examine the behaviour of electron waves in a nonlinear environment and this is greatly facilitated by comparison with nonlinear hydrodynamic phenomena and the generation of non-dispersive soliton structures.

By this analogy the most convincing description of an activated valence electron that initiates chemical interaction, is as a flexible soliton structure. It enters the valence state on stimulation in a chemically crowded environment that decouples the valence electron from the field of the atomic core in the form of a nonlinear standing spherical wave.

The nonlinear wave structure of the electron also accounts for electron diffraction, the generation of X-rays and scattering phenomena, without a particle postulate. The popular notion of anomalous chemical properties arising from the relativistic motion of electron particles is firmly discounted.

Hydrodynamics

The difference between linear and nonlinear systems is illustrated by the hydrodynamic concepts of laminar and turbulent flow. The laminar flow state represents an ideal and unstable linear system that develops turbulence on the slightest disturbance, with the appearance of eddies and nonlinear behaviour.

In the theory of differential equations linear equations are of degree one, referring to the algebraic degree of the highest-order derivative in the equation. Mathematically, the difference between linear and nonlinear equations exists therein that any two solutions of a linear equation can be added together to form a new solution. In the case of nonlinear equations superposition fails.

As in hydrodynamics, nonlinear waves are non-oscillatory and develop when the wave velocity becomes amplitude dependent. A wave that breaks on approaching the beach is a typical example. The profile of Hokusai's famous great breaking wave is closely approximated by a golden logarithmic spiral.

The superposition of linear waves of slightly different wavelength leads to the formation of *wave packets*, constructs, once considered as models for matter waves. However, linear wave packets are *dispersive* and dissipate spontaneously. For this reason both de Broglie and Schrödinger mooted the possibility of eventually reformulating a wave theory for matter waves in terms of nonlinear waves.

It was in fact shown by Schrödinger in one of his original papers that the harmonic-oscillator wave functions over a narrow range of high quantum numbers describe a well-defined wave packet that moves between the oscillation limits with group velocity corresponding to a classical particle velocity and energy. The wave group remains compact and of constant extension, but the breadth and number of elementary wavelets vary with time, becoming narrower and more numerous as the centre of oscillation is approached, and smoothed out at the turning points; hence a function of velocity. This behaviour represents the first recognition of what is now known as quantum soliton behaviour.

Solitary waves were first observed in a shallow narrow canal by the Scottish engineer John Scott Russell in 1834. He noticed how the wave that was generated by the motion of a horse-drawn barge kept on moving as the boat came to a sudden stop. He followed this unusual wave on horseback for a long distance and subsequently managed to generate and study similar waves on other canals and experimental tanks. One point of interest was the nondispersive nature of the solitary wave as it moved over the surface of the water without disturbance.

A mathematical model to account for Scott Russell's observation was published years later by Korteweg and de Vries, two Dutch scientists who derived a differential equation that governs the propagation of waves along the surface of a narrow canal. It consists of nonlinear and dispersion terms that describe opposing effects. When these terms are in balance the equation describes a solitary wave that moves without change in shape. It shows how a nonlinear wave, about to break, combines with a dispersive wave, about to dissipate, to generate a persistent solitary wave.

It was shown that the equation has a solution of the form that correctly describes a hump-like shape as described by Scott Russell and moving at constant velocity without change in shape. As the velocity is amplitude dependent it follows that a taller wave moves faster than a small one.

Solitons are now known to occur in many forms. The most spectacular example is the Red Spot in the surface of the planet Jupiter. It represents a coherent structure as wide as the distance from the earth to the moon. Nonlinear surface waves in the Andaman Sea, observed from orbiting satellites, are nearly 150 km wide. The tsunamis generated by earthquakes are of the same type.

Pursuing the hydrodynamic analogy a little further it is noted how a deep-water wave that enters a shallow narrow canal is transformed into a non-dispersive solitary wave, known as a soliton. Non-linear distortion of linear plane waves, especially by computer simulation, confirms that the appearance of solitons is a general phenomenon and an attractive model of 'elementary particles'. In the case of an electron wave this happens when it approaches an impenetrable obstacle or encounters non-linear modulation in a cloud chamber.

The formation of a soliton can be understood as the balance between linear dispersion and nonlinear cresting of a wave packet.

The resulting soliton is persistent and moves without change in shape. It is described by a linear differential equation with an added nonlinear term.

Matter Waves

The notion of defining matter in wave formalism follows directly from the reformulation of Newton's laws of motion in terms of a minimum-action principle. In its most advanced form the mechanical action ($S = Et$), also known as Hamilton's principal function, describes material motion in terms of a moving surface with the same properties as a propagating wavefront. It takes the form of what is known as the Hamilton-Jacobi differential equation.[1]

1

$$\frac{\partial S}{\partial t} + \frac{(\nabla S)^2}{2m} + V = 0\,.$$

In order to incorporate de Broglie's proposal of matter with wave properties Schrödinger modified the general wave equation into a form[2] that resembles the HJ equation, on incorporating the quantum conditions, $E = h\nu$ and $p = h/\lambda$, that link mechanical variables to wave variables through the quantum of action, h.

The two equations both describe moving matter in wave formalism, starting from opposite points of view—particle mechanics *vs* wave mechanics. To demonstrate the formal relationship between the two schemes the postulated wave description of matter, in the form of an amplitude-phase function,[3] is substituted into Schrödinger's equation. On simplification the real part reduces into a form[4] that differs from the HJ equation only in the appearance of an extra term[5] that contains Planck's action constant, h, and is known as the quantum potential.

The formal resemblance between classical and non-classical equations is reassuring, but because of V_q there is a discontinuity that separates the two descriptions of the same system. Quantum theorists have agonized over this classical-quantum limit for ages. The problem is overcome by modification of Schrödinger's equation to read:

$$ih\frac{\partial\Psi}{\partial t} = \left(-\frac{\hbar^2}{2m}\nabla^2 + V + V_q\right)\Psi\,.$$

Substitution of $\Psi = Re^{iS/\hbar}$, as before, now leads directly to the HJ equation.

The modified equation is one form of what is known as a nonlinear Schrödinger equation (NLS), which transforms smoothly into the classical HJ equation with increasing mass of the system. Because of its nonlinearity solutions of the NLS are not linearly superimpossible as routinely used in traditional quantum chemistry.

2

$$i\hbar\frac{\partial\Psi}{\partial t} = \left(-\frac{\hbar^2}{2m}\nabla^2 + V\right)\Psi.$$

3

$$\Psi = Re^{iS/\hbar}\,.$$

4

$$\frac{\partial S}{\partial t} + \frac{(\nabla S)^2}{2m} - \frac{\hbar^2\nabla^2 R}{2mR} + V = 0.$$

5

$$V_q = -\frac{\hbar^2(\nabla^2 R)}{2mR}.$$

Nonlinear Chemical Phenomena

The failure of quantum mechanics to produce a credible theory of chemistry need no longer cause any surprise. As a theory in three dimensions it is inadequate to account for spin, the major non-classical attribute of non-classical objects. By separating time from the space variables the four-dimensional harmonic function that accounts for four-dimensional action is decomposed, being restricted to axial rotation (orbital angular momentum) and linear oscillation in three dimensions. Chemical interaction, which only occurs as a result of spin action, remains a theoretical mystery. This observation dramatically invalidates Barrow's prescription [1] to reproduce non-linear phenomena by starting from 'simple linear problems'.

Crippled by the removal of one of its crutches Schrödinger wave mechanics limps along as a semi-classical theory of chemistry that relies on the linear superposition of real wave functions. This option completely ignores the importance of non-linear effects, to the point of seriously contemplating the reality of half-dead cats. The mere fact that the curved space-time of general relativity is intrinsically nonlinear shows that this approach is doomed to failure.

The superficial resemblance of some wave-mechanical predictions to empirical chemical reality has inspired a major programme to enforce consilience by introduction, in the Barrow mode, of a myriad of correction factors, skillfully disguised as legitimate quantum effects [2]. However, not being able to produce meaningful non-empirical results, this enterprise is loosing credibility and fast running out of steam.

At the beginning of the twentieth century chemists had a working model of chemical affinity for the identification of all major types of chemical reaction and the atomic composition of most chemical compounds, with a real prospect of understanding the nature of chemical cohesion. Discovery of the electron and the three-dimensional shape of molecules, against the background of the periodic table of the elements, soon led to the development also of a working model of chemical interaction as an electrostatic effect. Still in common use, the *Lewis* model of chemical interaction was based on an atomic model in which a heavy nucleus was surrounded by an incomplete cubic array of electronic point particles.

At about the same time the half-forgotten Balmer formula of atomic spectra coupled with the first observation of quantum effects provided the stimulus for renewed interest in the detailed structure of atoms.

Any atomic model that invokes orbiting electrons has the fatal weakness of being at variance with the laws of electrodynamics, which stipulate that an accelerated charge on a circular orbit must radiate away its orbital energy and collapse into the nucleus. Efforts to sidestep this stipulation resulted in the postulate of special quantum-stabilized radiationless stationary states. This proposal that gained widespread support for more than a decade was one of the first mysterious 'quantum effects' that still plague the theory.

An ingenious proposal to rationalize the quantum atom proposed by Bohr, developed from an effort to derive the details of the electromagnetic field from curved space-time, in the same way that Einstein formulated the gravitational field. It was shown by Herman Weyl that allowing for a change of scale (gauge) on displacement of a vector in space-time, sufficient new variables, in addition to those that characterize gravity, were generated to account for the electromagnetic potentials. It was soon realized that the new scheme could not account for the appearance of sharp spectral lines. The proposal was shelved until it was modified by Schrödinger [3] to describe the effect as a change of complex phase rather than gauge. This means that if an electron is described by a complex field Ψ, its charge q, of density $\rho = |\Psi|^2$, remains invariant[6] under a phase transformation such as $\Psi \rightarrow \Psi' = \Psi \exp(iq\alpha)$.

An electron endowed with such phase variation was shown to satisfy the Bohr conditions without the radiation problem. The difference is that a complex phase cannot reasonably be associated with a particle, implying a wave model for the electron.

Another possible way out was to develop a system of quantum mechanics without reference to the physical reality of space and time. It is indeed possible to construct a mathematical framework that accounts for atomic spectra without reference to the orbital motion of electrons. The rationale behind this scheme, based on the algebra of matrix multiplication, was to claim that it only relies on measurable properties of atomic systems and not on the details of what goes on within an atom—details that are not directly measurable and therefore intrinsically unknown.

The quantum theory of physics developed from the elaborate, so-called Copenhagen scheme, based on this positivistic postulate. To get away from any reality assumption the concept of probability density is introduced with the stipulation that the observation of an electronic point particle is an artefact of the measurement procedure. For further details of this orthodoxy the interested reader is referred to the countless books on quantum physics. The topic is not considered any further in this discussion, except for pointing out that as a theory of chemistry it is next to useless.

The major need of chemistry is not to find the details of atomic structure so much as formulating a credible mechanism in terms of which to understand the nature of chemical change. By no conceivable mechanism could atoms of the Bohr type interact to form molecules and, despite some seductive mathematical models, the same is true of quantum-mechanical atoms exchanging electronic particles.

The need is to establish how neutral atoms become activated by environmental factors into a reactive state. A valuable clue is provided by theoretical studies of a hydrogen atom subjected to isotropic compression [4–6]. The problem is analyzed by imposing a boundary condition that restricts electronic motion to a finite sphere, on the differential wave equation for hydrogen. The simulated compression energizes the electron to the point of zero potential energy where it becomes decoupled from the nucleus and confined to a sphere of critical radius $r_0 = 1.835a_0$ (a_0 =Bohr radius).

---

[6] $\Psi^*\Psi e^{i\phi}e^{-i\phi} = |\Psi|^2(\cos\phi + i\sin\phi)(\cos\phi - i\sin\phi) = |\Psi|^2$.

The behaviour of an electron in this nonlinear state prompted Sommerfeld [4] to remark that the situation was 'in need of further investigation'. It no longer obeys the linear wave equation and the simulation of covalent interaction by the superposition of ground-state wave functions is a crude approximation at best.

The various energy sub-levels are affected differently by compression. Increased pressure eventually inverts the normal sequence in such a way that the sub-level with the largest quantum number l has the lowest energy.

Extending the investigation to non-hydrogen atoms by computer methods [7] confirmed the hydrogenic pattern and resulted in a periodic set of ionization radii (r_0) for all elements. Any atom, compressed to the critical ionization level, consists of an atomic core surrounded by a spherically enclosed freed electron with non-zero energy; that can only be interpreted as quantum-potential energy.

The valence state of an activated atom in a chemically crowded environment should be closely modelled by such a critically compressed atom. The valence electron, decoupled from the nucleus, is free to interact with another, similarly activated electron, and pull the parent cores together into a covalent link. The quantum (chemical) potential of the valence state provides a basic interpretation of the well-known electronegativity concept. On this basis the periodic ionization radii of the elements is anticipated to be of fundamental importance in the elucidation of covalent interactions.

Electron Waves

At this stage it becomes imperative to resolve the question of electron structure. Interaction between point electrons to initiate covalence remains as awkward as before, whereas constructive interference between standing waves is unproblematical.

A convincing resolution of the orbital-angular-momentum problem is suggested by the de Broglie wave model of an electron. In order to span any Bohr orbit without self-destruction an electron is pictured as a standing wave of wavelength $\lambda = 2\pi r/n$, where n is the quantum number. The postulated relationship between wavelength and momentum, $p = h/\lambda$, accounts for the quantum condition, $pr = nh/2\pi$, interpreted as the quantization of angular momentum before.

Because of an unfortunate historical accident, exactly the inverse argument was used by Einstein to explain the photoelectric effect and his reputation carried the day. In hindsight he could have invoked de Broglie's model to ascribe the effect to the interaction between two waves without the photon postulate and eliminate the notion of an electronic point particle at the same time. The tragicomic result was Einstein's later opposition to the probabilistic interpretation forcefully imposed upon the matter-wave solutions of Schrödinger's equation. In the event the unrealistic Copenhagen interpretation prevailed against the wave model, relegating a chemically meaningful atomic model to obscurity.

It is inferred from the theory of general relativity that elementary forms of matter appear as four-dimensional standing waves. With space and time in an entangled state the concept of wave motion in four dimensions is ill-defined. In three-dimensional space it is an illusion created by the separation of space and time variables. This interpretation is beautifully supported by Schrödinger's equations. The time-dependent form describes a product function of the type $\psi(x) \cdot e^{i\omega t}$, a running wave with a time-independent component, known as Schrödinger's amplitude function. This is the three-dimensional projection of four-dimensional matter waves. It is also the most useful form for simulating the properties of matter.

It is worth noting once more that temporal dynamic properties such as kinetic energy, velocity and angular momentum have any meaning in the amplitude equation only in association with the constant unit of action, $\hbar$, which is added on by de Broglie's matter-wave postulate. There is no evidence of moving or circulating charges, except in an applied magnetic field that couples to $\hbar$, as evidenced spectroscopically.

Solutions of Schrödinger's amplitude equation for H, in spherical polar coordinates, are product functions of three eigenvalue equations

$$\psi(r,\theta,\varphi) = R(r) \cdot Y(\theta,\varphi) = R(r) \cdot \Theta(\theta) \cdot \Phi(\varphi)$$

with quantum numbers $n = 1, 2 \ldots$; $l = 0 \ldots (n-1)$; $m_l = -l, \ldots 0 \ldots l$. The radial equation defines the principal energy levels and the spherical harmonics, Y, correlate with allowed sub-levels that formally resemble angular-momentum states. The holistic function $\psi(r,\theta,\varphi)$, best approximated by a spherical standing wave centred on the atomic nucleus, represents the electron.

Solution of the equation for non-hydrogen atoms has not been achieved and, by default, the one-electron solution is assumed to hold approximately for all atoms. In order to correlate the results with observed atomic spectra an energy ground state in one-to-one correspondence between the total quantized electronic charge and the lowest atomic levels, allowing for an additional spin quantum number of $m_s = \pm\frac{1}{2}$, is assumed. In an excited, or activated, state the charge is spread over levels of higher quantum number. There are no stationary states, or quantum jumps, no room for electronic point particles, and the quantum integers arise naturally from the nodal structure of the (hyper)-spherical wave.

The concept of stationary states, related by quantum jumps, is a remnant of Bohr's mechanical model that acquired new meaning in the separation of wave functions into energy and angular-momentum components. In its four-dimensional hypercomplex form such decomposition has no meaning. Even in three-dimensional projection the notion of electrons at specific energy levels or orbital states is meaningless.

More correctly, an electron fluid in the form of a spherical standing wave displays no orbital structure or energy levels. In its ground state it resembles a stretched guitar string. When plucked, the guitar string goes into vibration, as dictated by its allowed vibrational modes. A mild disturbance activates only low frequency modes and more violent disturbance activates higher-frequency overtones. Exactly the same happens to an atom. As an external agency activates the normal harmonic modes, energy is radiated at a level commensurate with the activated frequencies.

The spherical standing wave responds in its entirety, as a whole, in a way that depends on all four-dimensional modes, correctly described by so many integers and half-integers, called quantum numbers. Low energy disturbance activates low-frequency response, as before. More violent attack activates higher frequencies, up to the ionization limit. The penultimate stage corresponds to high-energy electron bombardment that stimulates the emission of X-rays. The so-called K-level X-rays are the equivalent of the Lyman series of hydrogen.

The relaxation of an activated atom, back to the ground state, occurs in cascading steps. Each step represents the relaxation of a high-frequency overtone to a lower-frequency mode, each of these characterized by an integer, or a set of integers. The energy produced in the process is well described by a Balmer-type formula.

The notion that X-rays are produced when an energetic electron penetrates into the core of an atom, where it knocks out the most tightly bound electron, to be replaced by another from the next energy level, generates more questions than answers, while the simple resonance effect proposed here avoids that complication.

The wave-mechanical atomic model implies a highly flexible wave structure for an electron. Independent evidence reveals an electronic beam to diffract like a plane wave and to impact on a fluorescent screen, or leave a cloud-chamber trace, like a stream of particles. The Copenhagen model rationalizes this behaviour by the truism that, depending on external circumstances, an electron behaves like either a wave or a particle. The wave model readily accounts for such behaviour as environmental, non-linear modification of the electron wave.

A free electron in the vacuum resembles an electromagnetic disturbance with Compton wavelength $\lambda_C = h/mc$. Subject to nonlinear perturbation in more crowded environments the electron propagates with de Broglie wavelength $\lambda_{dB} = h/mv$. In the hydrogen atom where $\lambda_{dB} = 2\pi a_0$, the ratio $\lambda_C/\lambda_{dB} = e^2/\hbar c = \alpha$, defines the fine-structure constant.

Propagation of a wave through a nonlinear medium, like the restriction of laminar flow in hydrodynamics, causes cresting of the wave, resulting in a wave train of longer wavelength.

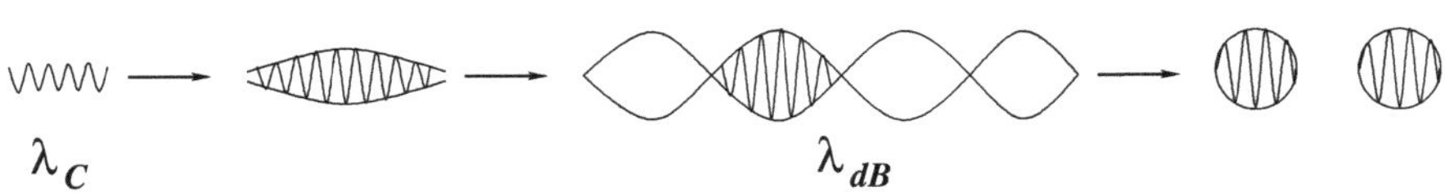

It is in this form that an electronic wave train with de Broglie wavelength that depends on electric field strength undergoes diffraction. In media at an increased level of nonlinearity, such as a Wilson cloud chamber, the wavetrain separates into individual solitons, each with the mass, charge and spin of an electron, and for all practical purposes appear to behave like particles of finite size in scattering experiments [8].

Non-Linear Interactions

Not only electrons, but also neutrons, protons and even molecules, up to the size of fullerenes, are known experimentally to show diffraction effects. The inescapable conclusion is that the formation of molecules depends on the merger of extranuclear electronic waves into flexible standing wave structures wrapped around the embedded nuclei. Intermolecular interactions arise from further overlap between molecular waves to form supramolecular assemblies or molecular crystals. All solid structures are eventually of this type.

Familiar molecular models are no more than a mapping of nuclear sites within the wave structure. Connectivity patterns have no more than geometrical meaning and do not represent 'chemical bonds'. The flexibility of homogeneous matter depends on the nature of individual atomic wave structures and the mass of the embedded nuclei. The nature of any wave structure, however complex, depends on the equilibrium redistribution of nodes and wave crests in four-dimensional space-time. The final pattern is expected to be a function of whole numbers and space-time topology. To unravel such patterns should be the objective of theoretical chemistry.

Matter waves interact not only among themselves but also with electromagnetic waves, which are essentially of the same type. This type of interaction can be studied experimentally by spectroscopic and diffraction techniques. It is in this regard that traditional wave mechanics became an indispensible tool in chemical analyses.

Only three models have provided meaningful insight into chemistry—the kinetic theory of ideal gases, periodic classification of the elements and wave mechanics. Although, having demonstrated that wave mechanics has in no way superseded the former 'classical' theories, it cannot be discounted, being an important advance, albeit as an approximation. The dilemma is that wave mechanics is traditionally handled as a subset of the much wider quantum theory with all of its baggage, found objectionable in chemical context. Resolution of this dilemma is not seen to be effected by yet another interpretation of quantum theory, but rather by a clean break with the current paradigm, which is firmly based on the paper presented by Heisenberg and Born at the famous 1927 Solvay Conference, stating:

> We regard quantum mechanics as a complete theory for which the fundamental physical and mathematical hypotheses are no longer susceptible of modification.

However, problems have been identified with the mathematical formalism, with the probabilistic interpretation, with the role of observers (the moon is not there when nobody looks), with quantum uncertainty, with non-local interaction and, above all, with wave-particle duality.

The only way to disentangle this web is by cutting the Gordian knot. Not to produce two incompatible worlds that meet in a mysterious unidentified limit, but two complementary approaches to analyze the same reality. On the one hand there is the Hamilton-Jacobi formulation of mechanical systems that resembles the propagation of a wave front. On the other hand the Schrödinger-de Broglie equation describes three-dimensional waves that simulate the motion of mechanical objects. At the macroscopic level the wave nature of matter is effectively masked and HJ mechanics provides an adequate description. At the microscopic level the wave model is crucial, with the inevitable appearance of integers that feature in the specification of wavelength and frequency instead of the continuous dynamic variables of momentum and energy. The introduction of quanta and point particles is a superfluous complication.

In order to avoid any further confusion it will be necessary to purge all chemical terminology of words such as quantum, particle, orbital, bond length, resonance and associated concepts. These concepts are perhaps less objectionable in physics, but for chemistry it is imperative to reformulate a working theory by refinement of the classical concepts of chemical affinity.

To enable such a reformulation the first step is to elucidate the difference between what became known as classical and non-classical phenomena. The spontaneous response would be that a non-classical system is invariably recognized in the appearance of Planck's constant, which is supposed to signal the breakdown of classical Newtonian mechanics. However, this is a delusion. Newton's laws, which correctly underpin particle mechanics, had been known for ages as unsuitable to describe wave motion. The distinction between particle and wave motion was first realized on discovery of optical interference and diffraction effects, not analyzable by the methods of geometrical optics. In the same way that the wave model of Huygens saved the day for optics, the Schrödinger-de Broglie wave model succeeded to elucidate sub-atomic physics where particle mechanics does not succeed.

At this stage the only useful distinction to be made is between particle mechanics and wave mechanics. These models cannot possibly apply at the same time. The relativistic model of four-dimensionally curved space-time is interpreted to define ponderable matter and energy, coining Einstein's description, as *condensations of space-time*, with an assumed wavelike structure. What is traditionally described as non-classical systems will now be described wave-mechanically and all macroscopic systems by particle mechanics.

On making this distinction it is essential to note that the term *wave mechanics* used here is not necessarily identical with its traditional use that incorporates the particle concept. Henceforth no attempt will be made to simulate wave motion in terms of a probabilistic particle model. In chemical systems this stipulation refers specifically to electrons. Any particle-like behaviour is a manifestation of non-linearity.

In the wave-mechanical analysis of molecular spectra it was noted that the potential-energy term which had been added into Schrödinger's equation, corresponded exactly to the 'classical' potential energy of particle mechanics, whereas the kinetic energy is 'quantized'. This assumption creates a conceptual problem in the case of the quantum harmonic oscillator where potential and kinetic energies are continually interconverted. The process seems to require an operator that interconverts continuous and discrete variables.

An assumption to get around this problem is that a quantum-mechanical system within a classical potential field acquires quantum-potential energy, V_q, as defined before. Only this component interconverts with the kinetic energy. As pointed out before the addition of V_q turns the wave equation into a non-linear differential equation. As the oscillator moves in linear mode through its point of minimum displacement, kinetic energy is progressively converted into quantum potential energy with increasing nonlinearity.

Even more revealing, at a deeper level, is the demonstration [9] that the Rydberg-Ritz equation that relates wave number to atomic number,

$$\bar{\nu} = R\left(\frac{1}{n^2} - \frac{1}{k^2}\right)$$

as a linear graphical relationship with the same slope and intercept,[7] is not identically valid. The most accurate spectroscopic measurements of the twentieth century, sponsored by IUPAC, reveal a statistically meaningful consistent difference between R_s and R_i. The original Balmer conjecture, which antedates this equation, and is universally accepted as the irreducible foundation of quantum mechanics, is thus demonstrated to be no more than a good approximation.

The rigorous equality of R_s and R_i in the Balmer formula, is intimately linked to the assumption of stationary states and quantum jumps. In the wave model of atomic spectra such a stipulation does not occur and the observed variability is to be expected. The simple fact that observed spectral lines are not infinitely sharp, denies the existence of stationary states. Complicated calculations to rationalize finite linewidths are finally all based on an uncertainty principle. We repeat that such calculations invariably consist of efforts to simulate wave motion in terms of probabilistic particle mechanics. The wave model accounts directly for all spectroscopic observations.

It is instructive to note that Niels Bohr himself had reservations about the Copenhagen programme as the ultimate interpretation of quantum mechanics, as evidenced by his perception of atomic periodicity [10]:

> ... the atomic number may be said to signify an important step towards the solution of a problem which for a long time has been one of the boldest dreams of natural science, namely to build up an understanding of the regularities of nature upon the consideration of pure numbers.

We can go one step further by noting that the ubiquitous whole numbers of atomic physics and spectroscopy are associated exclusively with wave structures and should go a long way towards the elucidation of atomic regularities without the need of solving complicated differential equations.

---

[7]For the Lyman series, $n = 1$, the plot of $\bar{\nu}$ vs $-1/k^2$ has the equation

$$\bar{\nu} = -R/k^2 + R\,, \quad \text{of the form,} \quad y = mx + c\,, \quad \text{the straight line with} \quad m = c.$$

This approach signals the parting of the ways for theoretical physics and chemistry, starting from common premises. Theoretical physics should explore the solution of non-linear differential equations in four dimensions by computerized numerical techniques, while chemistry should concentrate on the analysis of Bohr's natural system.

Unfortunately the perceived mysteries of quantum mechanics and relativity have a strange fascination for chemists. Such rapture is highlighted by the popular pursuit of some theoretical chemists to examine the effects of relativity theory in chemistry. One favourite topic is to trace the yellow colour of gold metal, in contrast to silver, to the relativistic motion of its *s* and *p* valence electrons.[8] Another is to account for the low melting point of mercury and the voltage of the lead-acid battery [11].

Apart from elaborate hand-waving and reference to Einstein and Dirac, as the bottom line it is stated that, because of relativistic mass increase the speed of an electron on a Bohr orbit increases to such an extent that:

> ... the valence s and p electrons have high speeds in the inner parts (in other words, orthogonality constraints against core states) ...

artfully dodging the more common terminology of penetrating orbitals.

However, the effect remains the same: The relativistic argument may be plausible, but the quantum model is untenable. The very concept of electrons speeding along quantum orbits was rejected a hundred years ago, together with Bohr's model. Like a charged object in a synchrotron, an orbiting or penetrating electronic particle must radiate away its energy as it changes direction. Hiding behind wave-particle duality is of no avail in this case. In Bohm's causal interpretation [12] of the concept any electron in a stationary state is predicted to have zero kinetic energy. It was clearly stated by Dirac in 1936 [13]:

> ... that the present quantum mechanics, with its conservation of energy and momentum, forms a satisfactory theory only when applied non-relativistically, to problems involving small velocities, and loses most of its generality and beauty when one attempts to make it relativistic. ... we see the need for a profound alteration in current theoretical ideas, involving a departure from the conservation laws, before we can hope to get a satisfactory relativistic quantum mechanics.

It is not relativity *per se* that affects chemical phenomena, but the four-dimensional nature of the electron density function, which cannot be simulated by the two-dimensional Bohr model.

The wave nature and diffraction properties of elementary matter provide an obvious way of introducing the ubiquitous whole numbers commonly ascribed to ill-defined quantum effects. Number theory is an attractive vehicle for this pursuit and has been demonstrated [14] to account empirically for many chemical phenomena.

The convincing demonstration [9] that the Rydberg-Ritz formula implies complex rather than real quantum numbers sounds a warning that number theory of

[8] Without mentioning the red colour of copper.

higher order (e.g. quaternion or octonion) could be the ultimate mathematical model for theoretical chemistry. However, this prospect lies strictly in the future.

References

1. J.D. Barrow, *Theories of Everything* (Vintage, London, 1992)
2. J.C.A. Boeyens, C.J.H. Schutte, Assumptions of quantum chemistry, in [15], 153–180 (2012)
3. E. Schrödinger, Eine bemerkenswerte Eigenschaft der Quantenbahnen eines einzelnen Elektrons. Z. Phys. **12**, 13–23 (1922)
4. A. Sommerfeld, H. Welker, Künstliche Grenzbedingungen beim Keplerproblem. Ann. Phys. **32**, 56–65 (1938)
5. S.R. de Groot, C.A. ten Seldam, On the energy levels of a model of the compressed hydrogen atom. Physica **12**, 669–682 (1946)
6. S. Goldman, C. Joslin, Spectroscopic properties of an isotropically compressed hydrogen atom. J. Phys. Chem. **96**, 6021–6027 (1992)
7. J.C.A. Boeyens, Ionization radii of compressed atoms. J. Chem. Soc. Faraday Trans. **90**, 3377–3381 (1994)
8. J.C.A. Boeyens, *The Chemistry of Matter Waves* (Springer, Dordrecht, 2013)
9. C.J.H. Schutte, Is the Rydberg-Ritz relationship valid? Struct. Bond **148**, 49–69 (2013)
10. N. Bohr, *Atomic Theory and the Description of Nature* (Univ. Press, Cambridge, 1934)
11. P. Pyykkö, Relativistic effects in chemistry. Ann. Rev. Phys. Chem. **63**, 45–64 (2012); Relativistic effects in structural chemistry. Chem Rev. **88**, 563–594 1988
12. P.R. Holland, *The Quantum Theory of Motion* (University Press, Cambridge, 1993)
13. P.A.M. Dirac, Does conservation of energy hold in atomic processes? Nature, 298–299 (1936)
14. J.C.A. Boeyens, P. Comba (eds.), *Electronic Structure and Number Theory* (Springer, Heidelberg, 2013)
15. M.V. Putz, *Chemical Information and Computational Challenges in the 21st Century* (Nova, New York, 2012)

Chapter 8
Space, Time and Matter

The realization that unification of quantum and relativity theories can only be achieved under four-dimensional formulation resolves most of the vexing quantum aberrations, without explaining the assumed nonlinearity of quantum effects. The remarkable effectiveness of wave mechanics in the elucidation of all quantum phenomena argues convincingly for a linear theory. However, the collapse or reduction of a wave function, and even quantum jumps, are nonlinear events, not governed by the wave equation; an inference carefully avoided by quantum philosophers. Primas [1] summarized a common perception in the words:

> ... that consciousness modifies the usual laws of physics and that the action of the mind on matter is associated with a grossly nonlinear time evolution equation for the state vector. Since the introduction of *ad hoc* modifications can ruin any theory, the proposals of nonlinear time evolutions for quantal states have no power to convince.

What an amazing cop-out. Why does the *ad hoc* assumption of quantum jumps and wave-function collapse not ruin the Copenhagen theory? However, without the complicating factor of consciousness, dragged into the argument by 'reduction of the wave packet', the conclusion on nonlinearity is in total accord with those reached in the previous chapter. Only the origin of a nonlinear term needs to be clarified and verification is not too hard to find in the non-Euclidean geometry of general relativity.

Cosmologists distinguish between an underlying curved space-time and the local Euclidean tangent space of common experience. Physical laws formulated in tangent space are linear approximations of the nonlinear laws of general relativity. That is why the concept of curved space-time is hard to swallow: the local tangent environment appears and responds as a Euclidean construct.

All of the philosophical speculation about space, time and the vacuum amounts to groping for an understanding of the origin and nature of matter. It is agreed fairly generally, but not universally, that matter is the product of a curved aether, according to the general theory of relativity. The form in which elementary matter emerges is

J.C.A. Boeyens, *The Quantum Gamble*, DOI 10.1007/978-3-319-41621-2_8

more contentious. The theories of particle physics favour a model of point particles endowed with mass, due to the Higgs mechanism and symmetry breaking. The alternative scenario of matter waves that occur as distortions of the curved aether results in a different model of elementary matter.

It will be argued that the way in which to understand elementary matter is by re-interpretation of the data that underpin the physics theory of elementary particles, but in terms of standing waves. A possible limitation exists therein that the theory of quantum chromodynamics that underpins the standard model is by orders more complicated than Schrödinger wave mechanics.

Wave mechanics was used in the previous two chapters as the starting point to address the problems of theoretical chemistry in terms of nonlinear hyper-spherical waves and solitons, instead of point particles. This approach avoids the complications that arise from the linear superposition of wave functions and the fanciful interpretation of quantum measurements that requires the collapse of a wave packet by intelligent intervention. It eliminates the awkward concepts of wave-particle duality and of a classical-quantum limit. In four-dimensional formalism, which emphasizes the holistic nature of spinors, Bell's analysis of non-local quantum events becomes irrelevant and the exclusion principle acquires physical meaning.

Theoretical physics is plagued by many of the same inconsistencies and could benefit by contemplating the same remedy. It could be of decisive importance in elementary-particle physics and its symbiosis with cosmology. All interactions in QCD are of the same type as Newtonian gravitation—action at a distance, mediated by unobservable virtual agents. The entire theory relies on speculative fields with exclusion principles and broken gauge symmetries. The all-important concept of symmetry breaking is shown in an appendix to simply reflect the difference between global and local phase invariance, which refers to tangent Euclidean space and non-Euclidean Riemannian space-time respectively. It has no cosmological meaning.

It will be argued that in terms a standing-wave model of elementary matter interactions are readily modelled as constructive interference in four-dimensional space-time, coding for characteristic values of mass, charge and spin. It predicts an internal wave structure for baryons with the same three-fold symmetry, assumed in the standard quark model.

The existence, or non-existence, of antimatter is analyzed in relation to space-time involution. Towards this end a short introduction to projective geometry is provided in an appendix. An attempt is made to demystify the power of mathematics in physical theory and the meaning of motion and time in a four-dimensional world.

Particle Physics

The standard model of particle physics, which is intimately linked to big-bang cosmology, is the ultimate in point-particle theories (atomism). It posits the appearance of matter as arising from the interaction between 'dark matter' and the 'Higgs field'. To date there is no physical evidence for the existence of either of these items, but in theory they represent more than 90 % of the (unobservable) universe. Dark matter is the sole product of the big bang and, together with 'dark energy', generate the dominant interactions (gravity and antigravity) that prevent the cosmos from either imploding or flying apart.

The Higgs field, first generated by 'cosmic inflation' was responsible for transforming some dark matter into a form, observable as massive point particles, which represent the basis of observational cosmology. As in particle physics, big-bang cosmology also recognizes the importance of antigravity, and both disciplines describe it quantitatively by a parameter, known as the *cosmological constant*, Λ.

Because of theoretical subtleties associated with the formulation of elusive parameters such as cosmic inflation, naked singularities, dark-matter haloes, event horizons, spontaneous symmetry breaking, flatness problem, and other imponderables, there is no agreement on the absolute value of the cosmological constant, invoked by particle-physicists and cosmologists alike. Cosmological theory requires a value close to zero, whereas particle physics needs $\Lambda > 10^{100}$ (a googol) in order to balance the books. This discrepancy does not seem to inhibit the common theory unduly.

The theory of elementary particles starts with the common forms of matter that occur within atoms, *viz.* electrons and nucleons. The electron is unproblematical, existing as a low-mass, negatively charged point particle, with intrinsic angular momentum that operates as a spin doublet. Its positively charged antagonist, known as a positron, occurs in antimatter with the same mass, but with antispin.

Like the electron, the nucleon is assumed to occur as a doublet, known as proton and neutron. Although these forms have the same spin, they are assumed to differ in *isospin*, with projections of $I_3 = \pm\frac{1}{2}$, to defeat the exclusion principle. Unlike ordinary spin, isospin operates in a different abstract (isospin) state. In line with this symmetry a proton can transform itself into a neutron, or *vice versa*. An awkward feature of isospin symmetry is to account for the $p-n$ mass difference and asymmetric electromagnetic effects. These anomalies imply that 'the degeneracy of the isospin states is only approximate' [2]; which seems to defeat the isospin theory.

The symmetry argument is stretched to its limit on contemplating the spontaneous decay of a free neutron into proton, electron and neutrino. The way out is found in the "creation" of the electron-neutrino pair by an unknown mechanism, said to be dictated by the 'Pauli principle' [2]:

> It may be that the neutron [...] finds a place in an orbital of lower energy which was occupied by protons, a state in which it could not have been when it was a proton before the decay.

If this logic appears somewhat daunting, there is worse to come. Scattering and collision studies, beyond the realm of atomic physics, reveal the existence of unanticipated sub-atomic debris, of the type shown here in the typical cloud-chamber track. There is no need to emphasize the complexity of such a data set. Each line, curve or squiggle is considered to carry information pertaining to the mass, charge and spin of a particle created in the event. All of this needs to be linked together in a self-consistent scheme.

Should the parameters associated with the known species fail to account for all detail it becomes necessary to postulate the appearance of previously unknown particles with new attributes beyond mass, charge and spin, not encountered or anticipated before. Careful re-assessment of the situation seemed to imply the existence of what became known as the 'zoo' of elementary sub-atomic particles.

A proposed scheme to rationalize such findings resulted in the formulation of the modern elementary-particle theory, exclusively based on symmetry models. The final assessment was that the complexity of hadron phenomena is due to the fact that hadrons are composite structures built up from a small number of structureless objects, called *quarks*. This is pure conjecture, but by now formulated in such elaborate detail as to be elevated to the level of revealed truth.

Each quark is a spin-$\frac{1}{2}$ point particle. Each baryon consists of three quarks, each of baryon number, $B = \frac{1}{3}$. In order to cover all angles five types of quark with appropriate quantum numbers of I_3 (isospin), S (strangeness) and C (charm) are required. Up, down, strange and charmed quarks carry charges of $\frac{2}{3}$, $-\frac{1}{3}$, $-\frac{1}{3}$, and $\frac{2}{3}$, respectively. In addition there are the flavoured bottom and top quarks. According to this scheme a proton consists of 2 up and 1 down quark and a neutron of 2 down and 1 up.

The standard model is not a physical theory, but more like an extensive game of SODUKO. The range of 'quantum numbers' is established as 'the most economical' and in order to complete the necessary input data set for further manipulation, extra quantum numbers may be added. Unlike those encountered in wave mechanics these quantum numbers do not represent eigenvalue solutions of some differential equation.

Quarks by themselves are powerless to synthesize hadrons unless operated on by the strong interaction, which 'is coupled only to the quark color'. The nature of strong interaction is summarized succinctly as [2]:

> [...]*the interaction between quarks and the strong field must produce a strong bond in color singlets, whereas color states other than singlets must acquire a high (possibly infinite) mass.* [...] this would account for the difficulty (impossibility?) of isolating quarks, because they are members of a noninvariant color triplet. The existence of a free quark would then require a high (perhaps infinite) energy.

> The recognition that color-invariant states are of such singular importance suggests that color invariance is a symmetry of nature. The field theory based on this symmetry, and on color as a source of fields, is called *quantum chromodynamics* (QCD).

The field quantum of the strong field is called a *gluon*, an unobserved massless gauge boson that binds (glues) quarks together in baryons and mesons in an interaction that generates the entire mass of the hadron. Quark mass, which appears to obey no logic, is ignored for all practical purposes.

One refrains from following the QCD procedure any further as it cascades into quantum uncertainties, Higgs fields, and intergluon interactions. It is hard to escape the impression that QCD attempts the impossible; to explain the nature of matter by ordering the chaotic evidence from cloud-chamber experiments in terms of non-interacting point particles endowed with mysterious quantum numbers in some abstract space. The quantum numbers come from nowhere and have no physical meaning. The gluon is perhaps the most complicated model ever constructed to explain a simple interaction between standing waves. The central argument of broken gauge symmetry is briefly outlined in Appendix C.

To highlight the enormity of the problem Jim Baggott [3] pointed out in summary how the theory

> - ... doesn't accommodate gravity.
> - It requires a collection of 61 'elementary' particles.
> - No clue can be gained from the masses themselves:
> - ... The top quark is about 134 times heavier than the charm quark, which is about 527 times heavier than the up quark. The bottom quark is about 41 times heavier than the strange quark, which is about 21 times heavier than the down quark. The down quarks (charge −1) are about 29 times heavier than the electron (charge −1);
> - You can keep looking at different combinations if you like, but there is no evidence of a pattern. No rhyme or reason.
>
> Breaking the symmetry to produce such divergent mass-energy scales appears to require an awful lot of fine-tuning.

However, critics of the standard model are few. The world has been mesmerized by the wizardry of particle physics. The rationale behind the theory sounds convincing: In the same way that the behaviour of extranuclear electrons is elucidated by the postulate of electron spin and the exclusion principle, the more complicated strong interaction is reduced to the interplay between the quantum numbers of the nuclear fields of isospin, strangeness, charm and colour, with matching exclusion principles. Still, there is one glaring difference: Electron spin is physically measurable and the exclusion postulate is empirically based on the observed structure of the periodic table of the elements. In comparison, the strong-field parameters are all figments of the imagination.

The most mysterious of the lot is the gluon. Its closest parallel in the history of science was known as *phlogiston*, the elusive alchemical principle of combustibility. In its most refined, final form it flaunted negative mass. The massless gluon goes one better—it generates nucleon mass by interaction with massless quarks. How strange, how charming!

The gluon (or gluons) originated as a postulate in efforts to understand the 'particles' of intermediate mass between electron and proton, known as *mesons*, observed in cosmic-ray showers. By implicating the meson as the agent, responsible for the interconversion of protons and neutrons two vexing problems could be solved in one go. Defining the meson as a transient composite of elementary particles of matter and antimatter, held together by a gluon, the internal structure of the nucleon was also resolved as reflected in the interconversion of proton to neutron:

$$p(q_u q_u q_d) + \mu(q_d \overline{q}_u) \rightarrow n(q_u q_d q_d)$$

The up antiquark $\overline{q}_u$ on the meson annihilates one of the up quarks on the proton. This proposal gave birth, not only to the quark, but also to the interminable egg dance of particle physics.

To turn back the clock the only option is to return to Fermi's model of the weak interaction that describes the decay of a free neutron in terms of the experimentally observed process,

$$n \rightarrow p + e + \nu$$

in which the neutrino is postulated to carry away the compensating half-unit of spin. This process simply accounts for the appearance of mesons as excitations of composite electron-neutrino standing waves.

But, enough of that. A simpler, alternative theory could be based, not on particles such as quarks, gluons and virtual photons, but on fragmented waves, which represent the turbulence, created in experimental collision studies and cosmic-ray showers. The symmetry remains essentially the same, but instead of operating on particles it refers to composite constructively interfering standing waves. The description of a meson as composed of a quark and an antiquark is no more general than a composite electron-neutrino wave, or antiwave, and their excitations.

Unlike the standard model, the alternative matter-wave theory derives directly from the general theory of relativity. The proposed model is therefore complicated by its reliance on the essential description of the curved aether in terms of intrinsically nonlinear, non-Euclidean, four-dimensional space-time and the alternative topological modes thereby implied. This topic is briefly reviewed in Appendix D.

Elementary Matter

Apart from the Big Bang the only other theory that explains the existence of matter is the theory of general relativity. It postulates a universe, consisting of an all-pervading aether, curved into a closed system, which results in the appearance of periodic distortions like the ripples on a pond. In the words of Mendel Sachs [4]:

> . . . all objects are manifestations of a single underlying order.
> There are no separate 'things'—electrons, human beings, plants, stars, galaxies, . . . —these are all correlated modes of [a] single universe.

The field equations of general relativity have the symbolic form of balanced tensors:

$$G(\text{geometry}) = T(\text{matter})$$

Given the variables on the left, the matter variables on the right are determined from the solutions of the field equations, and *vice versa*. In line with the properties of the electromagnetic field and the theory of special relativity the G tensor is defined in four dimensions. It is of decisive importance to note that the existence of matter implies that $T \neq 0$, which excludes the possibility of a flat aether and demands a non-Euclidean metric tensor, $g_{\mu\nu}(x)$, to define G.

Critics of general relativity [5–7] ridicule the notion that a coordinate system, which consists of nothing, can be curved. That is not what the theory implies. It postulates a physically real universe, colloquially described as the aether, in the shape of a four-dimensionally curved entity. The mathematical description of this shape requires the use of non-Euclidean geometry and hypercomplex algebra.

Periodic distortions that occur in the surface of the aether, as standing waves, are the elementary building blocks of atomic matter. Any sub-atomic entity occurs as a four-dimensional standing wave with four orthogonally directed vibrational modes. A longitudinal mode is traditionally defined along Z, with transverse modes X and Y. An active (temporal) mode occurs along the time axis.

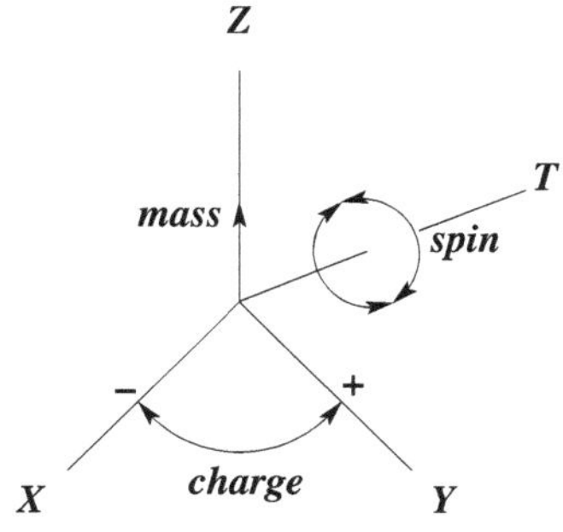

The nature of any elementary object depends on the dominant vibrational modes of its characteristic wave. A pure active mode represents spin in units of $\hbar/2$.

Transverse modes code for electric charge, $\pm e$, depending on the sense of spiral it entails along Z. Longitudinal modes impart mass. In leptonic matter longitudinal and transverse modes are well separated. Hadronic matter results from the linear combination of longitudinal and transverse modes to generate a spherical mode with three-fold internal symmetry [8]. The difference is modelled in the next figure, [9].

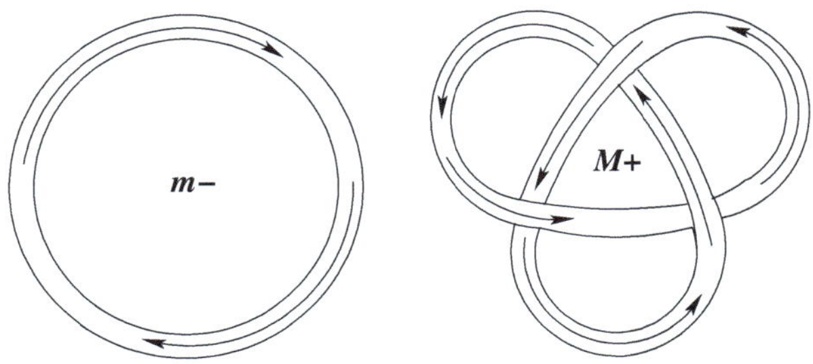

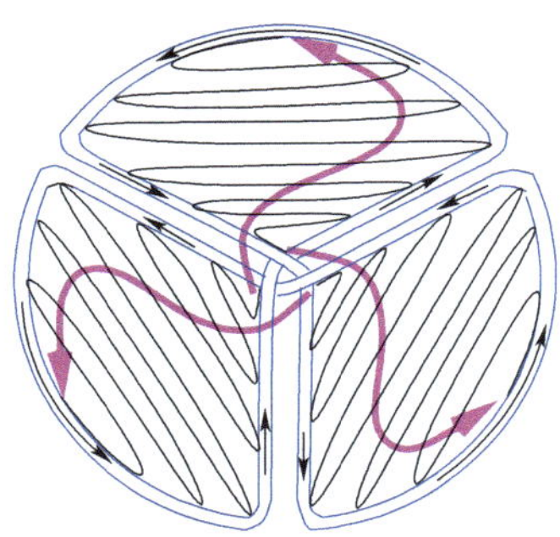

The internal transverse modes that code for electric charge, with a perpendicular magnetic mode, on an elementary standing wave are distinguished here by clockwise and anticlockwise spirals in the longitudinal direction, for negative and positive charges respectively. This model explains the concept considered to represent quarks as an internal symmetry mode of hadronic units, without independent existence. The interaction between quark-like standing waves to produce a proton in the form of a four-dimensional soliton is as shown schematically in the diagram.

The stable elementary forms of matter are known as neutrino, electron and proton. The neutrino is the most primitive form of 'matter'. It is massless and carries spin, but no charge, only responds to the so-called weak interaction and penetrates through kilometers of solid matter without deviation.

The way in which sub-atomic matter fuses into atoms depends on the internal wave structure of the elementary units.

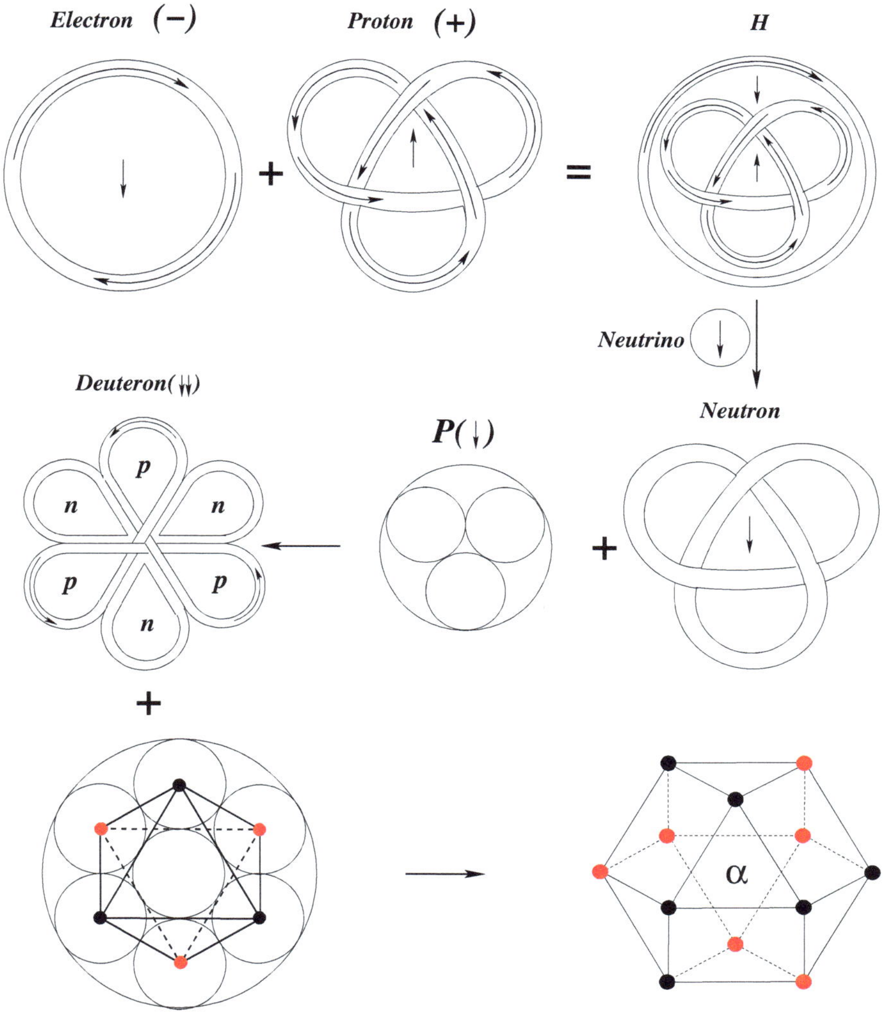

Electrostatic attraction between oppositely charged protons and electrons enables a fusion of the standing-wave structures to form hydrogen atoms. In a three-body interaction that also incorporates a neutrino, an electrically neutral hadron, or *neutron*, is produced. Its internal wave structure is complementary to that of a proton, such that proton and neutron waves can interpenetrate to form a stable new entity known as a *deuteron*. This intimate entanglement of standing waves is ascribed to a *strong force*.

The strong interaction is not confined to the proton-neutron pair, but is responsible for the stabilization of all atomic matter, as shown here schematically. It shows how interactions between electron, proton and neutrino lead to the formation of a hydrogen atom, a neutron, a deuteron with spin 1 and an alpha particle.

The strong interaction does not operate directly between protons without involving a neutron. The only stable nuclides with more protons than neutrons are ^{1}H and ^{3}He, the latter with relative abundance of 10^{-4}.

The standard symmetry model of elementary-particle physics postulates that a proton consists of two 'up' quarks with a charge of $2e/3$ each and a 'down' quark with $q = -e/3$, such that $q(uud) = +e$. A neutron with the complementary quark structure has $q = (ddu) = 0$. The up-down terminology refers to fermionic spin. To overcome the resulting exclusion problem for identical particles, an extra 'colour' quantum number is assigned to quarks.

A simpler way to avoid the problem is to assume the proton as consisting of a single standing wave with three entangled components, resulting in a holistic unit with unit positive charge and half a unit of spin. The concept of an elementary quark particle falls away. All interactions invoked here arise from constructive interference between standing waves and the involvement of gluons and virtual photons is not required.

The second way of representing the deuteron as an octahedral unit suggests the possibility of deriving the wave structures of more complicated nuclei from the known distribution of congruent kissing circles within a larger circle [10]. This way a nucleon resembles three confined circles and larger structures are as shown in the diagram.

Like the set of seven that defines ^{2}D, structures of ^{6}Li and ^{12}C project to optimally packed sets of 19 and 37 respectively. The packing of 61 spheres could be used to model ^{20}Ne. An incomplete set of 13 circles defines a cuboctahedron, proposed to represent an α-particle as the elementary unit, incorporated into all, more massive, nuclei. The set of 9 circles in a square models a tricapped trigonal prism, shown in tilted perspective to represent either ^{3}H or ^{3}He.

These diagrams reflect the correct composition and the relative orientation of elementary wave packets, but are not acceptable as models of atomic nuclei. Instead of distributing the groups of three wave packets that define protons and neutrons in the surface of a sphere, a more convincing picture is obtained by pulling them into close contact around the centre of the sphere. The composite waves of proton-neutron pairs become entangled as shown before schematically for a deuteron, and become further entangled in strong interaction with neighbouring deuterons. The proposal is illustrated by the structure assumed for an α-particle. It is noted that on contraction the wave packets isolated along the periphery define a second entangled pair. Such interactions result in all cases in the formation of hard spherical cores. To be demonstrated later on, the mode in which elementary wave packets stack up, results in a positive surface layer that interferes constructively with an enveloping spherical electronic wave.

The zoo of particle physics represents excited states of the stable forms. Excitation does not follow a continuous pattern, as the short-lived excitations are limited by the nodal structure of standing waves.

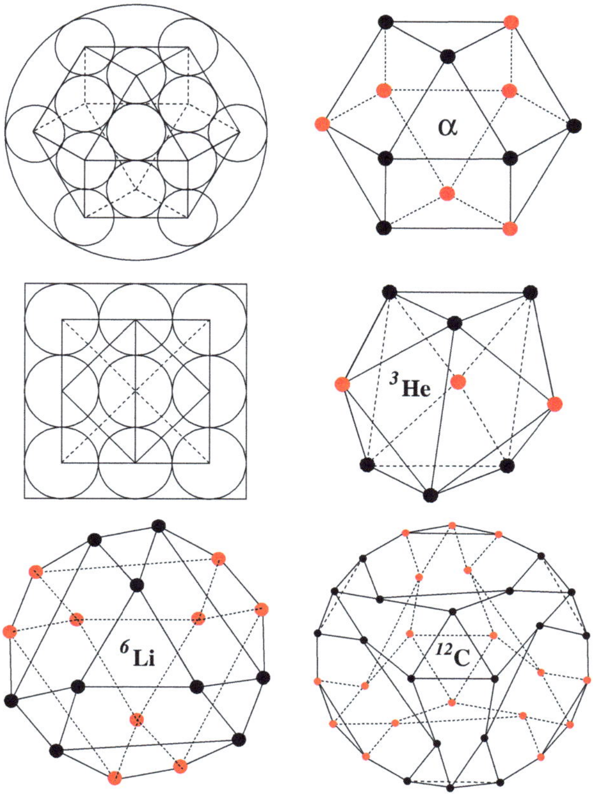

Each of the vibrational states that represent a characteristic form of matter has a chiral structure and the associated mirror images constitute antimatter. The only achiral entities are electromagnetic waves. They are considered to be racemic mixtures of spin and charge because they lack a longitudinal mode.

Particle *vs* Wave

In order to appreciate the fundamental difference between the particle and wave descriptions of elementary matter the popular exposition of the particle model by Harald Fritsch [11], a pioneer of quark theory, is a useful guide. The book traces the theory back to 'quantum theory', summarized on the hand of Bohr's stationary orbits, quantum jumps, electrons 'rotating about their axis', a relativistic exclusion principle and the enigma of antimatter. To round off the introductory argument it resorts to the concepts of 'electromagnetic quanta called photons',[1] virtual photons—'a consequence of the uncertainty principle', 'a strength parameter called the coupling constant'; 'called the fine structure constant' for electromagnetic interaction with '*real* photons like electrons', which proceed along 'lines of force'.

The wave model of matter is diametrically opposed to each and every one of these 'principles'.

The Bohr model of the atom is generally conceded to have been superseded by wave mechanics, which is pointedly ignored in the book. It is after all the only scheme that managed to make some sense of Bohr's energy levels, quantum jumps and angular-momentum assumptions. The outdated interpretation of electron spin and the exclusion principle as arising from the axial rotation of a point particle at relativistic speed, explains nothing and has no basis in quantum theory. On the contrary the hypercomplex formulation of four-dimensional harmonics leads directly to the definition of spinors, which describe the non-classical action that embodies both electron spin and angular momentum as a single holistic attribute of matter waves.

The vacillation about antimatter reflects the need to rely heavily on the existence of antimatter, despite the argument that '... makes the creation of antimatter a rather complicated business'. On the other hand, the wave model, which demands involuted space-time topology, neatly accounts for the mystery 'that the transition from matter to antimatter should [...] entail a mirroring effect in space' [11].

The characterization of electromagnetic interaction in terms of 'virtual' and 'real' photons, coupled through a fine structure constant, described by the author's collaborator, Richard Feynman [12], as '...one of the greatest damn mysteries of physics ...' carries little conviction. In the wave model there are no photons and the fine-structure constant is a special case, for electromagnetic interaction, of a parameter that details the shape of matter-wave packets.

The postulated electromagnetic field 'lines of force' are just there. In the wave model they represent a disturbance in the vacuum aether generated by a rotating, electrically charged wave packet.

On turning to atomic nuclei the problem to confront is explaining the stability of the nucleus against the electrostatic repulsion among the large number of protons, in terms of a 'strong interaction' which 'is active at very short distances'. Although it is stated that 'the only interaction of relevance to the atomic (electronic) cloud is electromagnetic interaction', the stability of the cloud of negative charges excites no further comment.

The strong interaction is rationalized, in the first instance, by Yukawa's proposed exchange of a heavy object between nucleons. The discovery in cosmic rays of intermediate mass objects, called *π-mesons*, was declared to provide experimental confirmation of Yukawa's model, now considered 'a quantum effect' [11]. In actual fact it is the same classical model proposed before by Newton to rationalize gravity

[1] 'The amount of energy carried by the photon is determined by the *wavelength of the light.*' This is the only reference to wave motion in the book.

as action at a distance, mediated by something which is not material. The π^+-meson with half-life of 2.6×10^{-8}s is, accordingly, probably best considered a 'virtual particle' in the Yukawa scheme.

Not only did a quantitative interpretation of the Yukawa model remain elusive, but the theoretical scene was further complicated by the experimental observation of many more 'elementary particles' in cosmic rays and on 'collision of stable particles'. The new particles were tentatively subdivided into hadrons, which respond to the strong force, and leptons that don't. Hadrons with non-integer spin are called baryons, different from hadronic mesons with integer spins.

All but one of the baryons are metastable and eventually decay into protons and non-hadronic material. This is interpreted to disclose a law that conserves the 'quantum' *baryon number*. This law clearly prevents the existence of 'any significant amount of antimatter in the universe'. Therefore the proton **must** eventually decay, even if it takes more than 10^{32}years. So far the search for a decaying proton has been fruitless.

The number of leptons, like hadrons, is not fixed and new leptons are occasionally observed in cosmic rays. At present three charged leptons and their associated neutrinos are known.

The standard theory of particle physics was inspired by the assumption that the neutron is just a symmetry modification of the proton. This assumption is based on the near-equality of proton and neutron masses, and despite the fact that the neutron is inherently unstable. The way in which both *nucleons* 'participate in the strong interaction' was interpreted as 'isospin invariance', despite the fact that 'isospin symmetry [is] not absolutely perfect'. The fateful conclusion that finally launched the standard theory is clearly revealed in the statement [11]:

> We suspect that isospin symmetry can be understood only by taking the quark structure into account.

Whereas baryons could be interpreted as three-quark systems, mesons are represented as quark-antiquark ($q\overline{q}$) systems. This is where the theory runs into its first major problem [11]:

> The neutral mesons [...] are slightly more complicated because it turns out that they are superpositions of several $\overline{q}q$ configurations. For the moment, let us not worry about this.

It is explained later on that:

> There is nothing wrong with this. In quantum mechanics we often deal with such mixtures.

However, these are not linear systems and this is not quantum mechanics.

The next problem (not to worry about?) is the difference in mass between hadrons with congruent quark structures. This discrepancy was rationalized by the truism that 'it takes energy to align the spins in the same direction as in a ρ meson'. Although quantitative calculations to support the supposition have never been done,[2] it is worth noting that in atomic spectroscopy spin inversion is treated as an excitation, which does not imply the creation of a new type of atom. Discounting excitations could therefore drastically deflate the particle zoo.

There is little point in following the standard theory as outlined by Fritsch [11] any further. Each new surprise in cosmic-ray or accelerator physics necessitates a modification of the theory. Starting out with one gluon per interaction it soon multiplied to eight and more. To quote [11]:

> ...just how many gluons are there inside a proton? It turns out that the number is rather large.

Like the gluons, the number of 'quantum numbers' also increases with each new development and so does the number of quarks. This number appears to have stabilized at six. And apparently one should not worry about the mass of a quark [11]:

> Of course, effective mass has nothing to do with the 'real' mass of the quarks, since quarks either do not exist or, if they do exist, are very heavy.

Eventually as the number of variables, quantum numbers, interactions and virtual particles gets out of hand the analysis reverts to group theory, finally ending up in QCD; the theory designed to render a three-quark structure antisymmetric, in line with the exclusion principle [11]:

> ...the Δ^{++} configuration should be antisymmetric with respect to the exchange of two quarks, and not symmetric as it seems to be. This discrepancy represents the major paradox of quark theory ...

The problem was overcome by inventing 'a new type of exclusion principle', based on colour. The three quarks in a baryon are assumed to be endowed with three different primary colours. However, the operational colour of each quark is *white*, resulting from a linear combination of the three primary colours. More than that, the linear combination that results 'in a sum of all terms [which] is antisymmetric with respect to the interchange of two quarks', requires six terms with alternating signs. Guided by symmetry-group theory quark configurations for any baryon, called *colour singlets*, can be selected 'to be totally antisymmetric with respect to colour'. This is not the end, but the beginning of QCD.

This quark story [11] ends in 1982 by mention of the DESY HERA electron-proton collider, anticipated to be commissioned in 1992. This facility remained operational until experiments were discontinued in 2007. The major achievement of its H1 and HERA experiments is highlighted on the DESY homepage, stating [13]:

> H1 and HERA discovered that the proton comprises not only three quarks, but rather a seething soup in which quarks, antiquarks and the gluons that act between them constantly appear and disappear.

[2] Spin inversion on a meson costs 600 MeV, whereas creation of a Δ 'particle' from a nucleon costs 300 MeV.

Perhaps there should be no conceptual problem in handling this situation with an expanded symmetry group, except that it might require a group of infinite dimensions.

In view of this new information it needs little imagination to now recognize the quark exposition as a vain effort to simulate the normal modes of a complex hyperspherical wave packet in terms of a particle model.

State of the Art

Given the failure of the standard model and the complications generated by *ad hoc* additions to mathematical quantum theory it is not too surprising to note the publication of a growing number of highly critical assessments of modern physics. Foremost among these is a treatise [14] by Lee Smolin, a disillusioned former afficionado, that severely criticizes the standard model of particle physics and string theory. Unzicker and Jones [15] are equally remorseless in their analysis of particle physics, string theory and cosmology—like Smolin, with devastating effect, but without suggesting any pragmatic resolution of the situation. Roychoudhuri [16] concentrates on optical phenomena and, based on a redefinition of 'the old concept of "ether" by the enhanced postulate of a complex tension field', recommends twelve ways of wielding Occam's razor to reduce the number of hypotheses that feature in modern physics.

All of these works differ from the approach outlined here in adhering to the concept of elementary, structureless point particles and unspecified topology of space-time. The three referred authors emphasize the same defects in the standard models of physics and cosmology, noted here, but remain unaware of the equally irrational conclusions of theoretical chemists, rationalized as quantum effects. The consensus, bordering on desperation, is reflected in the following statement [15]:

> True research is replaced by statistical kneading, which is supposed to legitimize the many parameters. And if despite all efforts, new contradictions show up, the reaction is as follows: If physicists do not understand the *what* of their theories, they'll introduce a new particle. If they don't understand the *when*, then it must have happened right after the Big Bang. If they don't understand the *where*, then of course it took place in an extra dimension. And if they don't understand the *how*, they will postulate a new interaction. If they don't understand the *how much*, a symmetry breaking will soon appear. If they don't understand anything, they will propose strings and branes. And if they loose interest in all understanding, there is always the strong anthropic principle. Things have come to a pretty pass.

In summary, the most pressing problems identified for resolution include:

- Cosmic matter-antimatter imbalance.
- Mystery of the fine-structure constant, α.
- Value of the gravitational constant, G.
- Meaning of spontaneous gauge-symmetry breaking.
- Background-independent equations of TGR.
- Noncommutative geometry.

- Modified Newtonian dynamics—dark matter and energy.
- Quantum nonlocality.
- Self-similar cosmic structure.
- Absolute CMB frame of reference.
- Explanation of spin.
- Black-hole singularity.

Most of these issues are readily resolved in terms of matter waves in projective space-time, as discussed in this work, albeit not always with quantitative rigour. Only the cosmological puzzles need further elucidation.

Most important of these are the background independence of the field equations of general relativity and the modified Newtonian dynamics, accounted for by dark energy and dark matter in standard cosmology. The first problem is that the fixed topology imposed by the standard model and by string theory does not allow for the dynamical structure that evolves in time, as required by TGR. However, this problem had been shown by Veblen [17] not to occur in the forgotten, projective, metric-free formulation of general relativity in which gauge invariance appears naturally.

The postulates of dark matter and energy resulted as an explanation of Zwicky's mysterious acceleration paradox, which points at the breakdown of Newton's laws at separations larger than the radius of the solar system. Any model based on uniformly curved space-time explains this anomaly to occur at a distance where the deviation between Euclidean tangent space and the underlying non-Euclidean geometry becomes substantial. This observation provides an opportunity to calculate the large-scale effective curvature of space-time.

The surprising realization that the cosmic microwave background represents an absolute frame of reference is no longer surprising in a closed universe where the CMB is no more than the standing black-body radiation of a closed cavity. It does not imply the existence of fixed points in a closed projective double cover.

Finally, the appearance of Newton's gravitational constant, G, in Einstein's equations correlates with the effective curvature of projective space-time in Veblen's analysis [17].

Mathematical Modelling

The elaborate group structure of the quark model sounds a word of warning against the overzealous use of mathematics to prescribe physical theory. This is not the same as finding the appropriate maths to recount the details of empirical conclusions. As a relevant example, it is becoming clear that there is no satisfactory explanation of the gravitational and electromagnetic fields in terms of three-dimensional vector theory. The minimum requirement appears to be a mathematical model based on four-dimensional hypercomplex algebra and non-Euclidean geometry. An inescapable consequence of this conclusion is a finite closed universe. As known from the theory of harmonic functions such a closed system invariably generates periodic phenomena. In three dimensions periodicity occurs either as one-dimensional vibrational modes or two-dimensional rotation. These modes can also be recognized as components of the four-dimensional quaternion, or spinor, mode. The latter is characteristic of the four-dimensional periodic objects that feature as elementary units of matter and energy.

It goes without saying that a mathematical description of elementary matter requires periodic algebra, the best known example of which is based on the counting numbers with a periodicity of unity between successive members of the set. The so-called Gaussian integers meet this requirement in the complex plane and Hurwitz integers in four-dimensions.

The title of Wigner's article [18],

> The unreasonable effectiveness of mathematics in the physical sciences,

has turned into one of the most frequently quoted phrases by philosophers of science. This aura of mystery largely disappears on noting the matching periodicities between the harmonics of closed systems and the natural numbers. It is conspicuous how quantum mechanics, when it models the existence of matter probabilistically rather than periodically, looses its predictive power, which is evident in spectroscopic analysis. Matter, as a product of space-time curvature, confirms that it consists of the same stuff as the aether, rather than some foreign material suspended in a void. It is of paramount importance to appreciate that flat Euclidean space-time is devoid of matter and energy. The normal modes of the closed non-Euclidean cover represent elementary units of matter and energy. When flattened out these modes disappear, as required by the field equations of general relativity.

To first approximation the normal modes appear as eigenvalues of a linear quaternionic action function in tangent space. This approximation breaks down at both astronomical and atomic scales. In the former instance the general curvature of space-time can no longer be ignored and at the atomic level the extreme concentration of matter implies severely curved local space-time. In particle physics the latter is compensated for by a local gauge transformation.

The action, or spin, function requires that elementary units of matter be characterized by 'internal periodic phenomena', in the terminology of de Broglie. These phenomena, which relate to mass, charge and spin, cause the interactions that produce the equilibrium forms of matter and radiation. On the planetary and astral scales matter has a noticeable effect on space-time curvature, recognized as local distortions. The general inertial motion of Newton's first law, coerced to follow the geodesic of local curvature, now appears as gravitational acceleration. Unlike the other forces of nature gravity therefore arises from local curving of the aether, giving rise to secondary gravitational waves of much lower intensity and more difficult to detect.

The only unexplained observation is the existence of antimatter, known to annihilate an equivalent quantity of matter on contact. All evidence points at the antimatter being chiral and related to matter as its mirror image; in topological terms an involuted form. In terms of projective space-time, antimatter then becomes antipodal to matter, within the octonionic double cover. Antipodal space-time sheets are separated by an interface, known as the vacuum, and characterized by opposite directions of time flow that prevent direct contact.

As explained before, a displacement along the interface, amounting to a rotation of 2π, causes chirality to invert and hence to gradually transform matter into antimatter. The inversion is not noticed by an observer that moves along the interface. The difference between matter and antimatter becomes apparent only on comparison of antipodal forms. The cosmological puzzle over the whereabouts of antimatter becomes a non-issue in that matter and antimatter now appear to be the same thing.

The Arrow of Time

The four-dimensional model proposed here leaves the vital concept of time unexplored. All scientific manipulations incorporates time as an independent linear variable, despite firm knowledge that any relativistic transformation of time invariably implies dependence on an allied transformation of space coordinates. The Lorentz transformation of special relativity demands

$$t' = \sqrt{1 - v^2/c^2} \cdot t$$

This is interpreted to mean that any measurement of space or time coordinates reflects a state of relative motion.

This conclusion leaves open the problem of how to interpret the irresistible flow (arrow) of time, experienced in a stationary system. The only conclusion is of an equally persistent, but imperceptible, steady motion through space. A compelling example of this is provided by the rotating universe predicted by Gödel's model that results from Einstein's equations on elimination of the absolute time coordinate [19].

Reformulated in projective topology [20] it is fully consistent with the model assumed here. The real projective plane has a topological fixed point only with respect to the transformation that defines a $2\pi n$ rotation as in the Gödel model. This fixed point can be placed anywhere and large-scale rotation must appear to be centred at the observer. The 'compass of inertia' that guides the rotation coincides with the involuted geodesic of high-dimensional projective geometry. To visualize the rotation, consider the motion of a point along the edge of a Möbius band and notice how it rotates around the central line without intersecting it.

Curvature of the projective surface occurs with respect to the local time coordinate of the non-Euclidean manifold, which is everywhere perpendicular to the surface. The interpenetrating dual spaces are separated in time, which prevents mutual annihilation. The same argument implies that the future is an approximate mirror image of the past. The sum total, which is manifest only in four dimensions, could imply an empty set, or a cyclic experience.

An observer, which is 'stationary' in space, remains oblivious of sharing the rotation and becomes aware of only the changing time component. This situation is not all that different from an observer who rotates, without noticing it, on the surface of the planet, but does register the arrow of time.

It is concluded that without motion there is no possible perception of time, which means that a state of absolute rest is meaningless. This paradoxical conclusion is just another manifestation of human inability to comprehend four-dimensional reality.

The widely accepted link between entropy and the arrow of time also acquires new meaning in four-dimensional analysis. It is implied that the second law of thermodynamics and the spontaneous increase of entropy also refer to four-dimensional effects, imperfectly detailed in three-dimensional projection into tangent Euclidean space. Like the arrow of time, entropy is implied to continuously change direction along a projective geodesic; and like matter, to have a conjugate form that occurs in equal measure on the cosmic scale. Over the complete range of 4π it integrates to zero.

The debate over the anomalous nature of entropy with respect to the time-reversible laws of physics acquires a satisfactory conclusion. Earlier speculation [9] on the 'quantum' nature of entropy now stands exposed at the same level as the postulates of quantum uncertainty and the measurement problem. However, the argument that time-reversal symmetry appears hidden in tangent Euclidean space, holds. Blurring of local world lines due to space-time curvature, which is ignored in the tangent approximation, are responsible for the perpetual motion in an ideal gas and the disorder ascribed to entropy [9].

The use of statistical quantum mechanics in an effort to account for the 'fuzziness' of classical chemical structures symbolizes the same problem and was no more than partially successful [21, 22]. It can now be stated with confidence that these fuzzy phenomena are not 'quantum effects', but manifestations of non-linearity, arising from four-dimensional space-time curvature.

The central problem relates to ill-suited extrapolation between tangent space and curved space-time. This is especially amiss in cosmological arguments that involve astronomical distances.

Denouement

Physical science has no meaning without a clear understanding of matter. The present discussion has reached the point where it confidently rejects atomism as a model and relies entirely on a wave description. The implied onus of accounting for the interaction between elementary entities without the participation of virtual objects such as photons, quarks and gluons, or the complications of chromodynamics, is seen as a major strength of the wave model at the same time.

Practical implementation of the scheme requires the formulation of an appropriate system of mechanics. The obvious possibility of Schrödinger wave mechanics does not suffice. A richer formalism of nonlinear matter-wave mechanics in non-

Euclidean, covariant, four-dimensional space-time is the minimum requirement. An essential feature of an appropriate theory must be a clear distinction between matter and antimatter, and between the recognized forces of Nature. The equivalence between matter and energy is implicit in the wave structure.

References

1. H. Primas, *Chemistry, Quantum Mechanics and Reductionism*, 2nd edn. (Springer, Berlin, 1983), p. 135
2. K. Gottfried, V.F. Weisskopf, *Concepts of Particle Physics* (Oxford University Press, Oxford, 1984)
3. J. Baggott, *Farewell to Reality* (Pegasus Books, New York, 2013)
4. M. Sachs, *Concepts of Modern Physics* (Imperial College Press, London, 2007)
5. N. Tesla, New York Herald Tribune, Sept. 11, 1932.
6. H. Arp, What has science come to? J. Scien. Explor. **14**, 447–454 (2000)
7. T. van Flandern, *Dark Matter, Missing Planets and Comets*, 2nd edn. (North Atlantic Books, Berkeley, 1999)
8. R.C. Jennison, Relativistic phase-locked cavities as particle models. J. Phys. A: Gen. **11**, 1525–1533 (1978)
9. J.C.A. Boeyens, *New Theories for Chemistry* (Elsevier, Amsterdam, 2005)
10. R.L. Graham, B.D. Lubaschevsky, K.J. Nurmela, P.R.J. Östergård, Dense packing of congruent circles in a circle. Discrete Math. **181**, 139–154 (1998)
11. H. Fritsch, *QUARKS: The Stuff of Matter* (Penguin Books, London, 1984)
12. R.P. Feynman, *QED* (Penguin Books, London, 1990)
13. www.desy.de/research/facilities_properties/hera/index_eng.html, Deutsches Elektronen-Synchroton (2015)
14. L. Smolin, *The Trouble with Physics* (Houghton mifflin, Boston, 2006)
15. A. Unzicker, S. Jones, *Bankrupting Physics* (Palgrave Macmillan, New York, 2013)
16. C. Roychoudhuri, *Causal Physics* (Baco Raton, 2014)
17. O. Veblen, *Projektive Relativitätstheorie*, Julius (Springer, Berlin, 1933). This monograph appears in English translation in [20]
18. E.P. Wigner, The unreasonable effectiveness of mathematics in the physical sciences. Comm. Pure. Appl. Math. **13**, 1–14 (1960)
19. K. Gödel, An example of a new type of cosmological solution of Einstein's field equations of gravity. Rev. Mod. Phys. **21**, 447–450 (1949)
20. J.C.A. Boeyens, *Chemical Cosmology*. Springer.com (2010)
21. A. Amann, Can quantum mechanics account for chemical structures? in [22], pp. 55–97
22. W. Gans, A. Amann, J.C.A. Boeyens, *Fundamental Principles of Molecular Modeling* (Plenum, New York, 1996)

Chapter 9
Matter-Wave Mechanics

The easy part of this investigation has now come to an end with the simple conclusion that traditional quantum theory, as a descriptor of reality, is incomplete. The more onerous responsibility that flows from this is to indicate the direction in which an alternative approach should develop in order to produce a theory of matter, consistent with the empirical observations of chemistry and atomic physics.

The most difficult part is to get away from the mechanical worldview, pioneered by Newton and systematically refined through successive phases from classical mechanics (CM), through quantum mechanics (QM), quantum electrodynamics (QED), quantum field theory (QFT) and culminating in quantum chromodynamics (QCD). The observational data of QCD, collected at the business end of an atom smasher, resemble the foam left on the beach by a breaking wave, more than anything else. The interpretation is as complex as unscrambling an egg. It could be more rewarding to examine the wave before it breaks.

The essential argument has remained the same for more than three hundred years: Matter consists of elementary point particles held together by mechanical interaction. Identity of the ultimate building blocks came to be recognized at progressively deeper levels of atom, nucleon, meson, electron and quark, but the essence has remained the same. The mode of interaction may have changed, but the mechanics stayed the same as reflected in units of energy.

To defuse the notion of action-at-a-distance, which Newton considered an 'absurdity', he declared [1] that

> ... it is inconceivable, that brute Matter should, *without the Mediation of something else, which is not material*, operate upon, and affect other Matter without mutual contact.

This notion has survived as either graviton, electron, virtual photon or gluon as mediators of the various *forces* of Nature. Only the electron, as mediator of covalent chemical interaction, has been successfully associated with a physically observable 'particle', albeit as a hybrid wave.

In order to discount this worldview at all levels it is necessary to dispose of the concepts of point particle and of virtual boson, in all its forms. Towards this end

J.C.A. Boeyens, *The Quantum Gamble*, DOI 10.1007/978-3-319-41621-2_9

it is necessary to reconsider the concept of mechanical motion in four-dimensional context and the way it gives rise to the emergence of matter.

The most primitive form of matter, the neutrino, is considered the most suitable example to elucidate the relationship between elementary and ponderable forms of matter as a general introduction to matter-wave mechanics.

A holistically structured universe will be accepted axiomatically as the seminal assumption that underpins all of natural philosophy, and as the logical starting point for this discussion. By this argument the origin of matter and energy can only be found in the vacuum; the ultimate bedrock of physical reality. As a guiding principle the theory of relativity is invoked. It stipulates that a consistent account of relative motion, pertaining to both mechanical motion and electromagnetic propagation, demands formulation in terms of four-dimensionally entangled space-time variables.

Motion in Four Dimensions

The geometry of special relativity was briefly discussed in the early chapters and visualized in two-dimensional Minkowski projection. Because of the complex

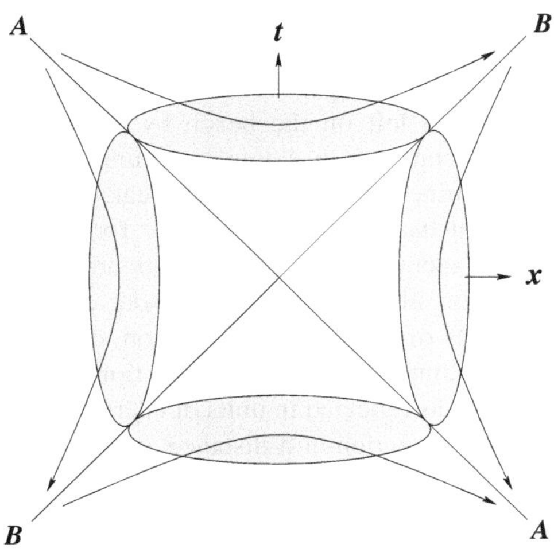

nature of the time coordinate the isotropic lines (*AA*) and (*BB*) have some unusual properties. Not only is the 'distance' between any two points on an isotropic line calculated as identically zero, but the line is also perpendicular to itself. The two isotropic lines are therefore one and the same and must join up geometrically. Assuming this to happen in a point at infinity each complete cycle $A - B - B - A$ or $B - A - A - B$ defines an involution as in the surface of a Möbius strip. The four-dimensional equivalent defines a *projective plane* which, like a Möbius band, is one-sided and non-orientable with a period of 4π, like the spin function. This conclusion has no logical understanding in three-dimensional Euclidean space. The common stipulation of Minkowski space-time as pseudo-Euclidean is therefore misleading and wrong. The only way to rationalize complex Minkowski space-time is by assuming that the isotropic surface has the topology of a closed projective plane. It follows that the notion of curved space-time is not an *assumption* of general relativity, but a natural consequence of special relativity. This is Occam's razor in action:

All things being equal, the simplest solution tends to be the best one.

The solution proposed here relies on the single assumption that mechanical and electromagnetic events obey the same covariant relativistic principle. The resulting formulation demands, not only four-dimensional space-time, but also closed topology, resulting from constant curvature with involution.

The implied involuted geometry shows that time and space cones in the diagram are separated by an interfacial surface between regions of inverted chirality. The double cover in four dimensions is enclosed on all sides by the projective interface and represents an attractive model for the physical universe. The intrinsic involution implies that Minkowski space-time is non-Euclidean. In the familiar three-dimensional world it appears to be Euclidean, topologically interpreted as the space, tangent to the underlying projectively curved space-time. Tangent space goes with Euclidean geometry and linear analysis, whereas curved space-time is intrinsically nonlinear. Minkowski space-time distinguishes between linear events in the surface of the light cones and material objects in time-like motion. The region traditionally labeled space-like is now more appropriately recognized as anti-timelike. The electromagnetic field is confined to the achiral interface, while gravity, which acts on fermions, operates in the double cover.

The linear interface may be likened to ripples in laminar flow on the surface of a pond. Curvature creates nonlinear turbulence that appears as ponderable matter in the double cover.

All debates on the interpretation of quantum mechanics must end in confusion, unless the quantum and classical models of the world are clearly distinguished. The classical model is based on the assumption that persistent fragmentation of matter terminates in a set of elementary particles that resist further subdivision, but retain the innate quality to predict the behaviour of matter in the bulk. A non-classical alternative starts at the other extreme with a featureless plenum that develops periodic wave structures in a topologically closed universe. These elementary waves coalesce into bigger units that exhibit all the known properties of ponderable matter.

Classical mechanics analyzes the interaction between particles, and non-classical mechanics studies the interaction between wave structures. The two models do not refer to classical and non-classical domains—they both model the same world, but from different points of view. It so happens that at different levels of aggregation one or the other provides a more convenient description. Attempts to describe classical structures non-classically, or *vice versa*, inevitably end up with illogical conclusions.

By the time that Schrödinger came to this same conclusion, the Copenhagen interpretation was too firmly entrenched for revision. He wrote [2] that a

> . . . finite universe is in itself the natural and wall-less box, which engenders atomicity by the necessary discreteness of its proper modes of vibration.

In a letter to Sommerfeld, quoted by [3] he elaborates:

> The world is finite and *therefore* it is atomistic—because a finite system possesses discrete proper frequencies. And in this way the general theory of relativity gives birth to quantum theory. That's the quintessence.[1]

He continues [2];

> . . . quantum mechanics has to abandon the idea of individuality in different particles of the same kind, two electrons for example. [. . .] This makes it fairly clear that one has to start by investigating the universe as a whole and to derive the methods, if any, for dealing with small isolated systems from the former investigation.

Schrödinger's alternative [4, 5] is in line with the four-dimensional model proposed here. The curvature of space-time is responsible for the appearance of matter and in order to describe it correctly it is necessary to respect the demands of relativity theory. This means that the mathematical description of elementary matter requires the use of hypercomplex, quaternion algebra, without the separation of space and time variables.

By considering the meaning of continuous matter density in space-time, Eddington [6, p. 147], arrived at the following conclusion:

> Density multiplied by volume in space gives us *mass* or, what appears to be the same thing, *energy*. But from our space-time point of view, a far more important thing is density multiplied by a four-dimensional volume of space and time; this is *action*. The multiplication by three dimensions gives mass or energy; and the fourth multiplication gives mass or energy multiplied by time. Action is thus mass multiplied by time, or energy multiplied by time, and is more fundamental than either.
> Action is the curvature of the world.

The earlier conclusion that the distortion of Euclidean space-time generates elementary units of matter may now be modified to state that elementary units of *action* occur as condensations in curved space-time. This means that general relativity implies, not only the appearance of a gravitational field, but also of a universal periodic field, characterized by $\hbar$. For lack of better terminology this is interpreted as the generation of four-dimensional waves in the vacuum.

Neutrino Physics

Of all known sub-atomic species the one that fits this description best is the *neutrino*. According to the standard model of particle physics neutrinos are similar to electrons, with half a unit of spin, do not carry electric charge and appear to be

[1] That means the aether.

massless, moving through the universe at the speed of light. Because neutrinos are only affected by the weak nuclear force that features at short range in radioactive interaction they have the ability to pass for great distances through matter without being affected by it. On rare occasions when neutrinos interact with nucleons they are transformed into electrons or other leptons. If this happens by interaction with a nucleon of an oxygen atom in a large volume of water the created electron moves through the medium faster than light and becomes a source of C̃erenkov radiation, which can be picked up by suitably placed photodetectors.

In the standard elementary-particle model a majority of neutrinos are assumed to date from the big bang. Alternatively they are the elementary standing waves characteristic of uniformly curved space-time. It is likely that in high concentration neutrinos and antineutrinos interact more readily, resulting in the creation of all different forms of fermionic matter, as well as bosons. In the equilibrium mix of the known universe they are diluted to a level at which they are largely inactive.

New findings in neutrino physics over the previous two decades are increasingly at variance with the standard model. A recent review [7], a research protocol from APS [8] and a popular review [9] have been consulted to prepare the following summary.

Some general conclusions include [9]:

> The Standard Model, as it stands, has no good explanation for why the Universe has anything in it at all, says Mark Messier, physics professor at Indiana University
> Neutrinos are at least a million times lighter than an electron, though no experiment has been able to definitely measure their mass.
> They travel extremely close to the speed of light.
> But perhaps the neutrino's strangest property is that they don't necessarily finish their travels with the same identity that they started it with.
> Somewhere along their journey from the Sun, they changed their type among their three flavors. This oscillation indicated they indeed have mass.
> "Every time we were able to measure a property of neutrinos, we were surprised by it", says Patrick Huber, a neutrino physicist and associate physics professor at Virginia Tech.
> ... neutrino flavors seem to be affected by regular matter as they flip by.
> Last month, Japanese experimenters demonstrated this oscillation effect by finding that neutrinos shine more brightly at night. As electron neutrinos stream from the Sun toward Earth, they oscillate to muon and tau neutrinos. But as they pass through the dense matter of our planet, some of them switch back. This suggests some quantum-mechanical transformation is taking place as the neutrinos interact with matter in the Earth, specifically its electrons.
> Because neutrinos are so much lighter than any other particle, the Higgs mechanism is unlikely to endow them with mass.
> Messier: "We see evidence that we don't have the full picture, that this nice story we tell ourselves about masses and mixings may be incomplete".

A number of unsolved problems, research objectives and important new insight had been highlighted [8]:

> Neutrinos, alone among matter particles, could be their own antiparticles. Are they?
> As time passes, or the neutrino travels, the quantum waves that accompany the different parts get out of step because the masses are different. Depending on the distance travelled, what was originally produced as an electron flavored 'neutrino' can become mu flavored or

> tau flavored as the components shift. This is the phenomenon called neutrino oscillations, and it provides our best evidence that neutrinos have distinct, nonzero masses.
> There is a lot still to learn about the masses and flavors. We are now trying to measure the flavor contents of each neutrino, and we represent them by 3 trigonometric angles called $\theta_{12'}$, $\theta_{13'}$ and $\theta_{23'}$. The masses themselves are only known within broad ranges, although oscillations tell us quite a lot about the differences.
> ... neutrinos changing flavor as they traveled through space was the only way to explain the missing 'atmospheric' neutrinos.
> In the Standard Model with massless neutrinos, it is not possible to generate the matter-antimatter asymmetry of the universe dynamically.

The statement that [8]:

> The particle and the antiparticle have identical mass and spin... Neutral particles are special in that they can be their own antiparticles. This is true of several neutral particles that are not fermions, including the photon—the particle of light.

is interpreted to imply that neutrinos can be their own antiparticles, an issue to be urgently investigated in the study of 'nuclear double β-decay'. However, there is still no experimental evidence that supports this statement [7, 10], while theoretically there is a clear distinction between spin and antispin [11], and [7]:

> According to $V - A$ theory the neutrino is left-handed and the antineutrino right-handed.

An interesting observation [7, 12] is that the assumed process of neutrino oscillation cannot be described by a particle model, which necessitates a 'wave-packet model', defined by a superposition of plane waves. This treatment is introduced in a footnote [7] without reference to the dispersion of linear wave packets, which invalidates the entire discussion. Even the issue of neutrino mass is thereby left open to question.

The pedantic statement about neutrinos that [9]:

> They are just inherently quantum mechanical ...

is supposed to justify the use of any argument that appears vaguely appropriate in support of some wild conclusion. However, [8]:

> There are many other deep physics principles that can be tested through neutrino studies. The discovery of effects such as the violation of Lorentz invariance, of the equivalence principle, or of CPT-invariance, to name only a few, would force us to redefine the basic tools—relativity, quantum mechanics—we use in order to describe Nature. Physics and astrophysics would be led to the very challenging but rewarding path of fundamental revision.

On the other hand it may be sufficient to simply abandon the particle concept at all levels of microphysics.

An intriguing proposal [13] seeks to relate neutrino oscillations to the golden ratio. Without going into broken-symmetry arguments [14] or the nature of neutrino mass and coupling matrices, the proposal makes intuitive sense as due to the non-linear response of fermionic waves to curved space-time. However, this argument requires more than oblique reference to wave packets [12] and serious consideration of neutrino waves.

One strategy to avoid the illogical features of Copenhagen quantum theory is by resorting to a wave model for elementary matter, until now considered only in the detail of simple sine waves. No attempt has been made to endow matter waves with fermionic half-integer spin. This becomes an important issue in the comparison of neutrino waves with electromagnetic radiation. Assuming, to first approximation, that neutrinos are massless, the crucial difference between the two types of wave is associated with their spins. The integer spin of an electromagnetic wave, consisting of two orthogonal transverse components, is described as polarization, which is directly measurable and readily accounted for. The fermionic spin carried by neutrinos implies a fundamentally different mode of polarization, experimentally evident in the weak interaction between neutrinos and ordinary (fermionic) matter.

To better appreciate the nature of neutrino waves it should be recalled that electron spin can only be described correctly in quaternion algebra. To recognize the neutrino it therefore becomes necessary to consider the full quaternionic solutions of Maxwell's equations as originally derived, rather than the vectorial formulation. An important feature of quaternionic waves is that they consist of both transverse and longitudinal components. This conclusion has been derived by several authors. The seminal analysis is due to Silberstein [15] who defined the electromagnetic field in terms of a quaternionic *bivector* or *biquaternion*.

Using this formulation the biquaternion was shown [16] to be an 8-dimensional invariant that defines the detailed Lorentz transformation in Minkowski space-time and includes a longitudinal electroscalar wave as a possible description of a de Broglie wave. Essentially the same result was demonstrated [17], starting from a single quaternion and also leading to a neutral scalar field arising from vacuum fluctuations, that could represent neutrino waves.

For electromagnetic waves, known to interact with electronic charge, the transverse components are clearly dominant. The more penetrating neutrino waves, presumed to interact with nuclear matter through the weak force, may then be assumed more longitudinal in nature.[2]

The deviant behaviour of neutrino waves is strikingly illustrated by the transactional interpretation [19] of electromagnetic interaction. Briefly explained [19]:

> ... the process ... can be thought of as the emitter sending out a probe wave in various allowed directions, seeking a transaction. An absorber, responding to one of these probe waves, sends a verifying wave back to the emitter verifying the transaction and arranging for the transfer of energy and momentum.

The so-called 'retarted' signal from the emitter is propagated into the future, as a function of time, while the 'advanced' response from the absorber is propagated into the past—a function of negative time, $f(-t)$. It is important to note that the general wave equation has both retarded and advanced solutions, although the latter is traditionally ignored as physically meaningless.

Accepting the advanced response as legitimate means that the return signal reaches the emitter at the exact moment of first transmission to establish an

[2]It is intriguing to note that Schrödinger toyed with the same idea [18].

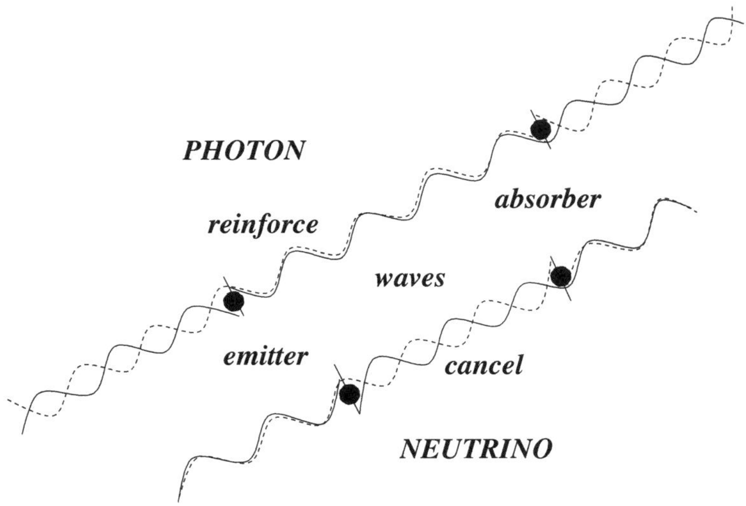

instantaneous[3] handshake. Assuming both retarded and advanced waves to undergo a phase shift on reaching their destination they interfere constructively to set up a standing wave between emitter and absorber. This standing wave that carries excess energy from emitter to absorber,corresponds to the photon of quantum theory.

The experimentally known behaviour of neutrinos, observed to penetrate the earth at the speed of light, without being absorbed, is clearly not of the same type. The diagram that models the neutrino, in its behaviour, as the exact opposite of the photon, implies that it never interacts with any matter. This is clearly not the case, as interaction is known to happen, albeit as a very rare event. Likewise, the transaction model, in predicting total absorption of all electromagnetic radiation, is also not valid without exception, as evidenced by the existence of the ubiquitous microwave background radiation.

Cramer [19] offers a cosmological explanation of neutrino behaviour. Assuming the space-time of an open universe to be Euclidean, he inserts a mirror for neutrino waves at the time of the big bang ($t = 0$) in order to validate the proposed transactional model.

In projectively closed cosmology the advanced and retarded neutrino waves are destined to join up through an involution. A transactional interaction between emitter and absorber would then differ from the assumed photon model only by an involution in the region where waves reinforce—i.e. an emitted neutrino is absorbed as an anti-neutrino. This scenario echoes the observation of particle physicists that equates a neutrino with its anti-particle.

The wave model goes a long way towards an explanation of neutrino oscillation.

[3]Recall the complex nature of a relativistic signal, equivalent to a standing wave in the surface of a Minkowski light cone.

The suggestions by Schrödinger that the structure and behaviour of electrons could be described in terms of a wave model caused a massive outcry among his contemporaries who favoured a point-particle description. It could be demonstrated that a wave-packet, constructed by linear superposition, propagates with a group velocity that matched the velocities derived from experimental observation of electrons, and accounted for the mysterious uncertainty principle of quantum particle mechanics. Although the argument was defeated by reference to the dispersion of wave packets it was smuggled into the Copenhagen model in the form of probability waves. The way to understand the logic of quantum mechanics is by noting that although the existence of wave packets are firmly denied on the one hand, a quantum measurement, on the other hand, is defined as the collapse of a wave packet.

The same technique came in handy as particle physics encountered the problem of neutrino oscillation, which had no explanation in the standard model. The problem was that experiments designed to count the number of solar neutrinos reaching the earth, observed only about one third of the predicted number. A partial resolution of the problem was provided by modification of the standard model to allow different non-zero masses for differently 'flavoured' neutrinos. Although suitable mathematical models could be formulated to account for the mass change of the neutrino in transit from the sun, the standard model does not provide a credible physical model for the 'oscillation'. However, there is a quantum explanation:—The neutrino turns into a flexible wave packet that adapts a form not observable by normal measurement techniques.

The problem is resolved miraculously by calling it a quantum event. So powerful is the theory that a concept found wanting almost a century ago can be revived with impunity in a time of need. The mere fact that the wave-packet description of an electron is taboo is irrelevant when faced with the solar neutrino problem.

Despite the self-contradiction the wave-packet argument has merit—in a wave theory of neutrinos, but not a probability-wave theory. Like acoustic waves, longitudinal neutrino waves should respond to a nonlinear environment by developing beats and soliton structures. It is conceivable for such waves to oscillate between different 'flavours' as they move from a vacuum into the planetary atmosphere and through the earth.

The observation of neutrino oscillation creates an awkward problem for the symmetry argument of elementary-particle physics. This theory distinguishes between electron, muon and tau neutrino particles to represent the symmetry, required to match the postulate of six quark particles. Instead, neutrino oscillation implies a single neutrino that freely changes its appearance.

Historical Note

It is generally agreed that the existence of the neutrino was postulated by Pauli in 1930 and experimentally confirmed only in 1956. However, from documented historical records [20–22] it is clear that neutrinos were observed and described by

Nikola Tesla as early as 1898. The term *neutrino* did not exist when Tesla made his discovery and his use of *cosmic ray* should also not be confused with its modern usage. Interpreting his conjectured superluminal speed simply as *high speed* the description makes more sense than most modern definitions of neutrinos.

In a letter to the New York Times Tesla wrote [21]:

> You have given considerable space to the subject of cosmic rays, which seems to have aroused general attention to an unusual degree. Inasmuch as I discovered this wonderful phenomenon and investigated it long before others began their researches, your readers may perhaps be interested in my own findings.
> The original idea was advanced and discussed by me in a series of articles on Röntgen rays and radioactivity, published from 1896 to 1898 in *The Electrical Review*.
> The experiments I undertook in 1896 were greatly facilitated through my invention of a novel form of vacuum tube suitable for operation by currents of many millions of volts and yielding effects of transcending intensities. This instrument has since been adopted by other investigators and most of the progress in several fields was achieved by its use.
> When radioactivity was discovered, it was thought to be an entirely new manifestation of energy limited to a few substances. I obtained sufficient evidence to convince me that such actions were general and in nature the same as those exhibited by my tubes. In these, minute corpuscles,[4] regarding which we are still in doubt, are shot from a highly electrified terminal against a target where they generate Röntgen or other rays by impact. Now, according to my theory, a radioactive body is simply a target which is continuously bombarded by infinitesimal bullets projected from all parts of the universe, and if this, then unknown, cosmic radiation could be wholly intercepted, radioactivity would cease.
> I made some progress in solving the mystery until in 1899 I obtained mathematical and experimental proofs that the sun and other heavenly bodies similarly conditioned emit rays of great energy which consist of inconceivably small particles animated by velocities vastly exceeding that of light. So great is the penetrative power of these rays that they can traverse thousands of miles of solid matter with but slight diminution of velocity.

Later on Tesla was reported by Alfred Albelli [22] as describing his experimental findings at Colorado Springs in 1899. He stated that:

> My findings are in disagreement with the theories more recently advanced.
> I have satisfied myself that the rays are not generated by the formation of new matter in space, a process which would be like water running uphill. According to my observations, they come from all the suns of the universe and in such abundance that the part contributed by our own sun is very insignificant by percentage. Some of these rays are of such terrific power that they can traverse through thousands of miles of solid matter.
>
> **Properties of Solar Rays**
>
> They have, furthermore, other extraordinary properties. This ray,[5] which I call the primary solar ray, gives rise to a secondary radiation by impact against the air and the cosmic dust scattered through space. If radium could be screened effectively against this ray it would cease to be radioactive.

Tesla's comment about the indifference of quantum physicists towards his work was highlighted in the original report of the interview, in the style reproduced below [22]:

---

[4]Electrons.

[5]Neutrino wave.

What About Today's Scientists?

"The scientists from Franklin to Morse were clear thinkers and did not produce erroneous theories. The scientists of today think deeply instead of clearly. One must be sane to think clearly, but one can think deeply and be quite insane.

"Today's scientists have substituted mathematics for experiments and they wander off through equation after equation and eventually build a structure which has no relation to reality."

—Nikola Tesla.

The difference between thinking clearly and thinking deeply is illustrated dramatically by the analysis of radioactive α-decay by Tesla and according to Gamow's tunnel theory, respectively.

The best calculation of a free α-particle's ($^4He^{2+}$) energy within an atomic nucleus is about 20 MeV short of the amount required to overcome the potential Coulomb barrier in order to escape from the nucleus.

According to Tesla, thinking clearly, all of the required energy is supplied by the absorption of a high-energy neutrino. "Not so", says Gamow, thinking deeply: "the α-particle is guided by a probability wave, with the mysterious ability to tunnel through the barrier and carry the α-particle with it to the outside".

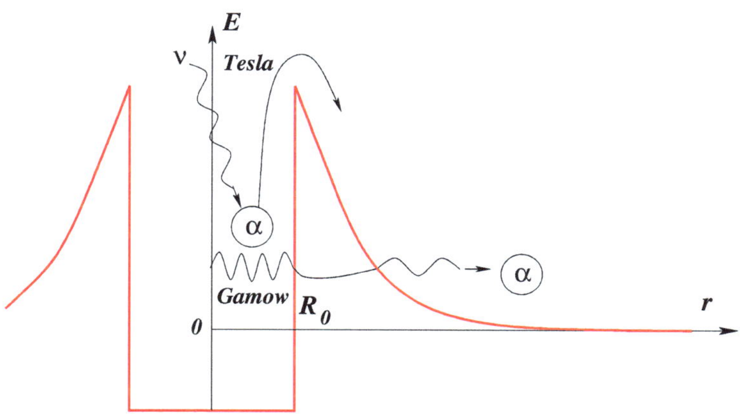

This process is contingent on the assumption that an α-particle is confined in the nuclear region only by electrostatic interaction. The way in which the α-particle came to be decoupled from the strong interaction that pervades all nucleons is explained [23] as follows:

> By borrowing energy for a short time according to Heisenberg's uncertainty principle, it will be able to travel beyond the range of the strong attractive forces of the remaining nucleons to a region where it is subject only to the electrical repulsion due to the protons ... the energy is borrowed according to the probabilistic laws of quantum theory ...

With such a generous patron available why not borrow all of the energy needed to escape in one go and bypass the hastle of tunneling? Be that as it may—Gamow's tunnel effect is just another of those clever concepts that explain nothing but prevent further analysis.

The same Gamow launched the big-bang theory as an April fool's joke. Like his cosmology, the concept of quantum-mechanical tunneling managed to establish itself as one of the pillars of modern physics. The joke is that the same quantum physicists who rejected de Broglie's proposal of a guiding wave that directs a quantum-mechanical particle in 1927, accepted Gamow's tunneling wave in 1928, with acclaim. Like the uncertainty principle the tunnel effect is extremely useful for avoiding the need of justifying illogical conclusions. It is routinely used to good effect by quantum chemists to rationalize sterically impossible intramolecular rearrangements.

The countless equations that specify the transition probabilities for particles over barriers serve no known useful purpose apart from annoying the likes of Tesla.

Back to Matter Waves

Next to the neutrino the most abundant form of elementary matter is the common lepton, known as an electron. On a microscale the most important conclusion to be drawn from quaternion physics relates to the structure of an electron. It is described by the quaternion eigenfunction of the four-dimensionally modified space-time Laplacian equation. Physically this is interpreted as a spinor, which in three-dimensional space projects as an object that spins in spherical mode. Any distortion of its natural spherical shape is governed by the three-dimensional spherical harmonics. In many-electron systems the symmetry of non-spherical distortion is traditionally interpreted as (orbital) angular momentum. This leads to the awkward situation of a hydrogenic electron with non-zero orbital angular momentum ($l > 0$), without a rotation axis ($m_l = 0$). Interpreted as a projection of four-dimensional action into 3-dimensional space, the contradiction disappears. It is implied that the decomposition of a quaternion spinor into complex and real components cannot reflect the situation holistically.

It is instructive to note that a multi-electron atom remains spherical in its ground state. This is confirmed by the fact that any degenerate set of Schrödinger eigenfunctions consists of a single real (spherical) function with $m_l = 0$ and conjugate pairs of complex functions with $m_l = \pm(0, l)$, which quench into spherical objects. In colloquial terms, an atomic nucleus is surrounded by a spherical uniform electric fluid with a total charge of Ze, rather than so many electrons. The fluid is quantized in the sense that it may be dispersed into Z identical units, but not beyond that. There are no sub-electronic electric charges. The fluid, like a single electron, consists of a standing wave made up of superimposed spherical waves of Compton wavelength.

In terms of persistent nonlinear wave structures the model becomes even more convincing. Any matter wave appears as a coarse wave with fine-grained internal structure. The wavelength ratio λ_C/λ_{dB}, describes the texture of the wave by a characteristic fine-structure parameter, in the customary terminology of Compton and de Broglie wavelengths.

For the hydrogen electron, considered to be guided by a wave of wavelength $2\pi a_0$, where $a_0 = \hbar^2/mc^2$ is the first Bohr radius, the parameter

$$\alpha^H = \frac{h}{mc} \cdot \frac{me^2}{2\pi\hbar^2} = \frac{e^2}{\hbar c},$$

is known as the fine-structure constant.

Making the same assumption for an electron of classical radius, $r_0 = e^2/mc^2$, the parameter appears inverted,

$$\alpha^c = \frac{h}{mc} \cdot \frac{mc^2}{2\pi e^2} = \frac{\hbar c}{e^2}.$$

Remarkably, $\alpha^c = 1/\alpha^H$.

An important clue to elucidate this result, seems to be missing. The classical electron appears to invert, inside out, when wrapped around the proton. The de Broglie model of an orbiting electron, guided by a one-dimensional wave makes no sense in this respect. More appropriately, the electron is seen to surround the proton in the form of a spherical standing wave with radial nodes at n^2a_0.

The electron of radius r_0 can only be a soliton and to account for the reciprocal relationship with a_0, *i.e.* $r_0/a_0 = \alpha^2$, it must somehow be associated with a superluminal speed, such as the phase velocity (v_ϕ) of the hydrogenic de Broglie wavepacket with group velocity v_g. It is well known that $v_g v_\phi = c^2$.

The same conclusion was reached by Schrödinger [24] who identified a superluminal component associated with electronic motion, ascribed to an interior non-linear high-frequency periodic trembling motion (Zitterbewegung). This is precisely the nonlinear term required to stabilize a nondispersive wave packet and identifies the classical radius as the ultimate limit to which an electronic soliton can be compressed.

It was surmised by Schrödinger that the phenomenon described here relates to electron spin, and this conjecture can now be supported by the nature of the four-dimensional spin function. Superluminal motion certainly makes more sense in three-dimensional projection of four-dimensional action. As emphasized before, a quaternionic spinor cannot be described correctly in 3-dimensional space.

Physics has the dilemma of irrefutable evidence for a four-dimensional world, but a genetic inability among physicists to visualize more than three dimensions. It is therefore not surprising to find that those instances in which reality is badly distorted in three-dimensional projection, inevitably lead to convoluted theories, bordering on the supernatural. Quantum mechanics is a prime example of such a theory. It was inspired by experimental results that defied explanation based on classical theory. It was first recognized in the study of microphysical systems, which in time came to be seen as deviating from the classical and therefore subject to a new theory, without relevance in macrophysics.

A more plausible interpretation is that the motion of ponderous objects, projected into tangent three-dimensional space, differs imperceptibly from four-dimensional reality in the local environment where a classical description suffices. It only becomes an issue for fast-moving objects and where particle mass approaches zero. The real meaning of both relativity and quantum theory is obscured by their formulation as alternatives to Newtonian mechanics that kick in at some classical limit.

The proliferation of elementary particles discovered by nuclear physicists presented a new challenge for quantum theory. At some stage it became evident that not all observations could represent new species and that many of these could be better described as excited states of stable particles. However, that would imply internal structure, not anticipated as the property of elementary, indivisible units of matter. Theories to account for the particle 'zoo' and their interactions are generally based on the gauge principle, reformulated by Weyl as a phase factor, and symmetry breaking.

The most conspicuous success of gauge theory occurred in the rationalization of the electromagnetic field. It was mentioned earlier how the conservation of electric charge is to be understood as invariance under a global phase transformation of the type $\Psi' = \Psi e^{i\alpha}$. This formulation is valid in tangent space, but not necessarily in the curved manifold of general relativity. In this case the gauge factor is no longer globally fixed, but could vary locally with the space-time coordinates at different points, $\alpha = f(x, t)$. This rules out local charge conservation, unless a compensating field that restores the invariance is generated in the same transformation. Straightforward mathematical analysis [25] shows this indeed to be the case.

The difference between Euclidean and curved space-time corresponds exactly to what is defined as 'spontaneous symmetry breaking' in elementary-particle physics. The only difference being that the high-symmetry state of particle physics is considered to refer to a false vacuum created by the big bang, and subsequently spontaneously broken in the process of cosmic inflation. Another complicating factor is presented by the need to associate phase with an elementary point particle, which requires the re-introduction of wave-particle duality.

The matter-wave model suggests an entirely different, simpler scenario. Like the electron, all sub-atomic matter is considered to be four-dimensional standing-wave structures, conveniently, but incorrectly, described as linear global gauge-invariant systems in Euclidean tangent space. Spontaneous symmetry breaking only means allowance for local variation of gauge parameters, with the appearance of appropriate compensating fields.

To account for interaction between elementary particles it was postulated to be mediated by virtual particles. When two electrons interact they exchange a virtual particle, or virtual photon, going in both directions at the same time. This may sound like a bizarre proposition, but perhaps not if it is a quantum effect. The uncertainty principle is assumed to allow the energy required to generate the virtual photon to be borrowed from somewhere, provided the deficit is made up within the allowed time span according to[6]:

$$\Delta E \Delta t \geq \hbar\,.$$

The good news is that energy conservation is preserved in the long run, but unfortunately the same is not true of momentum. The short-term loan of energy is provided by the quantum vacuum, drawing on the pool of zero-point energies. The momentum discrepancy occurs because virtual photons are 'off mass-shell' and therefore can exist in an unobservable intermediate state. As clear as mud.

Looking for an equivalent wave model to account for the interaction the detail of an electron in space-time is briefly revisited. An electron spinning in the vacuum moves without impediment but creates a periodic disturbance, like a propwash, in the environment. When the disturbances, created by a pair of electrons in close proximity, overlap, a standing wave with all the attributes ascribed to a virtual particle is formed. The standing wave communicates the nature of both wave packets, which decides the type of interaction that ensues.

The famous absorber theory of Wheeler and Feynman [26] may be read as a long-range extension of the same idea. In this model an excited-state electron (emitter) broadcasts a retarded signal (going forward in time) to which a potential absorber responds with an advanced reply (going backwards in time). The response reaches the emitter at the exact time of first emission to establish an instantaneous handshake in the form of a standing wave between emitter and absorber. This standing wave, the equivalent of a photon, does not move, but in case of an imbalance it carries the energy required for equilibrium from emitter to absorber at the speed of light.

The symmetry of the situation implies that, like an absorber which cannot function without an emitter, emission does not happen unless a suitable absorber is lined up. For the benefit of sceptical readers it is pointed out that the general wave equation is time symmetrical with both advanced and retarded solutions.

References

1. A.R. Hall, *The Rise of Modern Science, III.* (Collins, London, 1963), p. 316
2. E. Schrödinger, World structure. Nature **140**, 742–744 (1937)
3. A. Rüger, Atomism from cosmology: Erwin Schrödinger's work on wave mechanics and space–time structure. Hist. Stud. Phil. Sci. **18**, 377–401 (1988)
4. E. Schrödinger, The energy–momentum theorem for material waves. Ann. der Phys. (4), **82**, (1927): as translated in [5]

[6]Let $\Delta\nu$ denote the width of a spectral line corresponding to a range of energies ΔE. From $E = h\nu$ and the classical theory of harmonic resolving power follows $\Delta E = h\Delta\nu \geq h/\Delta t$, which is a statistical law involving many units and does not apply to the precision of individual measurements.

5. E. Schrödinger, *Collected Papers on Wave Mechanics*. Translated from 2nd German edition (Chelsea, New York, 1978)
6. A.S. Eddington, *Space, Time and Gravitation* (Univ. Press, Cambridge, 1921)
7. V. Barger, D. Marfatia, K. Whisnant, *The Physics of Neutrinos* (Princeton U.P., Princeton, 2012)
8. Members of the APS Multi-Divisional Neutrino Study, *The Neutrino Matrix*, November 2004. http://www.aps.org/policy/reports/multidivisional/neutrino/upload/main.pdf
9. R. Boyle, *Forget the Higgs* http://arstechnica.com/science/2014/04/forget-the-higgs-neutrinos-may-be-the-key-to-breaking-the-standard-model/
10. M. Goldhaber, L. Grodzins, A.W. Sunyar, Helicity of neutrinos. Phys. Rev. **109**, 1015–1017 (1958)
11. E.P. Batty-Pratt, T.J. Racey, Geometrical model for fundamental particles. Int. J. Theor. Phys. **19**, 437–475 (1980)
12. C. Giunti, Neutrino flavor states and the quantum theory of neutrino oscillations. AIP Conf. Proc **1029**, 3–19 (2008). [arXiv:0801.0653 [hep-ph]]
13. Y. Kajiyama, M. Raidal, A. Strumia, The golden ratio prediction for the solar neutrino mixing. Phys. Rev. D **76**, 117301 (2007)
14. F. Feruglio, A. Paris, The golden ratio prediction for the solar angle from a natural model with A_5 flavor symmetry. J. High Energy Phys. **2001**(3), article 101 (2011)
15. L. Silberstein, Quaternionic form of relativity. Phil. Mag. **23**, 790–809 (1912)
16. K.J. van Vlaenderen, A. Waser, Generalization of classical electrodynamics to admit a scalar field and longitudinal waves. Hadronic J. **24**, 609–628 (2001)
17. A.I. Arbab, Z.A. Satti, On the generalized Maxwell equations and their prediction of electroscalar wave. Prog. Phys. **2**, 8–13 (2009)
18. L. Bass, E. Schrödinger, Must the photon mass be zero? Proc. Roy. Soc. **232A**, 1–6 (1955)
19. J.G. Cramer, The arrow of electromagnetic time and generalized absorber theory. Found. Phys. **13**, 887–902 (1983)
20. J.T. Ratzlaff, *Tesla Said* (Tesla Book Co., Millbrae, Calif., 1934). ISBN 0-914119-00-1
21. N. Tesla, *Letter to the Editor*, New York Times, Feb. 6, 1932, P.16, col. 8. Reprinted in [20, p. 237]
22. N. Tesla, reported by A. Albelli, *Radio Power will Revolutionize the World*, Modern Mechanics and Inventions, July, 1934. Reprinted in [20, p. 272]
23. G.D. Coughlan, J.E. Dodd, *The Ideas of Particle Physics*, 2nd edn. (Cambridge U.P., Cambridge, 1991)
24. E. Schrödinger, *Zur Quantenmechanik des Elektrons*. Sitz. Ber. Preuss. Akad. Wiss. 63–72 (1931)
25. J.C.A. Boeyens, *The Chemistry of Matter Waves* (Springer, Dordrecht, 2013)
26. J.A. Wheeler, R.P. Feynman, Interaction with the absorber as the mechanism of radiation. Revs. Mod. Phys. **17**, 157–181 (1945)

Chapter 10
Epilogue

In conclusion it is necessary to point out that the purpose of this essay has not been to belittle the efforts of the quantum pioneers, but to caution against the construction of scientific theories by acclaim. By constant probing it is possible to expose possible flaws, even in the most reputable theory. That is the easy part.

Equally easy, is to add secondary weak assumptions to mask any defect, followed by another if necessary, and another ... The end product is a house of cards. The quantum theory of chemistry has reached the point, by following this procedure, where it is stated with serious conviction that "in the case of hybrid orbitals quantum numbers are not needed any more".

The more daunting approach is to challenge any flawed idea, even of the most respected authority. This is what scientists of the Enlightenment had to do with Aristotle's physics, what Copernicus and Galileo did to Ptolemaic cosmology, and what Lavoisier had to do about the phlogiston theory. In retrospect such actions are hailed as heroic, but in real time they caused endless animosity. Lavoisier ended up on the guillotine.

In the present instance understanding of the criticism against quantum theory is growing, but tempered by widespread reluctance to abandon a respected paradigm. A common response is that concepts such as sp^2 interaction have become part of chemical language and should therefore be retained without the orbital connotation. This is dangerous and unacceptable. It is argued that simulation by DFT methods produces useful results and should continue as part of *computational chemistry*, without invoking quantum theory. Fact is, that the methodology relies on expendable complications, inspired by suspect procedures, such as linear superposition, once considered essential as quantum techniques, but no longer valid.

Another respondent, focussed on career prospects in relation to research output, could see no benefit in switching to comparatively untried methods, and working without reliable software. To call this a mercenary attitude may be too harsh, but it is unscientific.

Others despair over giving up a user-friendly comfort zone in favour of obscure unfamiliar concepts such as space-time curvature, hypercomplex algebra,

J.C.A. Boeyens, *The Quantum Gamble*, DOI 10.1007/978-3-319-41621-2_10

self-similar symmetry, logarithmic spirals, gauge invariance and other outlandish things. These critics should remember how, at the time when Heisenberg announced the basis of quantum mechanics, he had never heard of a mathematical matrix, the concept that made him famous. To soften the blow the discussion is concluded by a brief summary of a logical argument that leads from ordinary common sense to a useful re-appraisal of physical science, that presumably avoids the illogical features of quantum mechanics, the gamble that never paid off.

On the paper trail through Tesla's utterances one encounters a statement that explains the widespread scepticism about the theory of relativity. He states [1]:

> I hold that space cannot be curved, for the simple reason that it can have no properties. It might as well be said that God has properties. He has not, but only attributes and these are of our own making. Of properties we can only speak when dealing with matter filling the space. To say that in the presence of large bodies space becomes curved, is equivalent to stating that something can act upon nothing.

In this instance it is not a question of *clear vs deep* thinking, but a simple matter of careless terminology.

To put the record straight, consider a finite four-dimensional physical aether, which is closed and therefore curved. It curves through an involution into a shape, topologically best described as the double cover of a projective plane. The interface is a vacuum devoid of all matter, which is concentrated as standing waves in the double cover. The interface is pervaded by achiral bosonic fields, with all fermions in the chiral double cover.

To account for the reciprocity between matter and curvature postulated by general relativity, it may be assumed that should the aether be flattened out into an infinite Euclidean plane it becomes completely featureless. All irregularities, recognized as matter and energy, as well as the interface, would disappear into a void. Without agonizing over the origin of the universe or the development of its assumed structure it is sufficient to postulate a balanced system in dynamic equilibrium. Popular arguments about the miraculous fine-tuning of physical constants serve no purpose.

A useful axiom is to assume that the stability and properties of the cosmos depend on its size. The state of equilibrium must be commensurate with maximum stability and optimal size. This way a unique cosmic structure is implied.

In this state all vital parameters are in fixed relationship, leaving no room for speculation about slightly different fine-structure constants or the speed of light. In its closed equilibrium state the size and curvature of the cosmos are invariant and in harmony with all properties that depend on these. 'Mysterious' fine-tuning goes together with the existence of the universe and leaves no room for teleological speculation. The only task of the cosmologist is to measure and interpret the parameters, characteristic of the observed structure.

One possible strategy is to postulate a trial structure for validation against all conceivable observations. The current investigation postulates a universe, closed with the topology of a real projective plane and general curvature to match a golden logarithmic spiral. Visualization of this four-dimensional structure, in the form of a

Boy's Surface

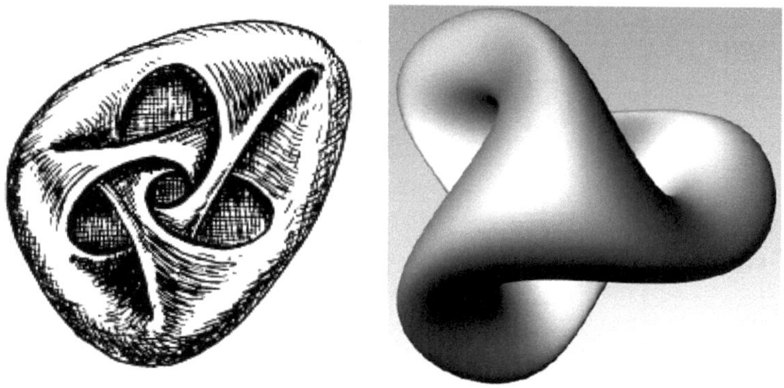

Flat view ***Round view***

three-dimensional shadow [2], is known as a Boy surface; it cuts through itself but has the advantage of having no singular points.

The immediate consequences are cyclic variation of space and time, finite size of the universe, self-similar symmetry throughout the cosmos, effective separation of matter and antimatter, increased distortion at long range of observations interpreted in tangent space, and the necessity to reconsider all cosmological models.

The most important consideration is that cosmic models should be consistent with mathematical practice, the most important of which is to be free of singularities. This is where standard cosmology is woefully inadequate. The entire edifice is based on the most glaring of singularities and stretches into an undefined infinity. It is littered with countless black-hole singularities and despite its infinite extent has no room for antimatter. The hogwash about naked and hidden singularities confuses rather than enlightens the argument. It simply has no credible mathematical basis.

The projective structure proposed here, does not lead to any known singularities. Black-hole singularity is eliminated by allowing discharge of infinitely dense matter through the cosmic interface. For each black hole there is an antipodal white hole. It is significant to note that the assumed double cover is in perfect register with octonion algebra as indicated before. At the same time analysis of the group operations on the projective plane has been shown [3] to be equivalent to quaternion multiplication.

The proposal of four-dimensional matter waves with spin, charge and mass is in line with both geometrical and algebraic demands of the assumed topology. All of the symmetries needed by particle physicists to balance the books, are the normal modes of four-dimensional waves in $2 \times 4 = 1 + 7$ dimensions of octonion algebra.

The re-interpretation of non-classical effects in terms of wave phenomena has serious implications beyond theoretical chemistry and physics. One of the most profound, concerns the physical size of the universe, seen as a closed rather than an infinite Euclidean structure. Another look at Boy's perception of a projective surface

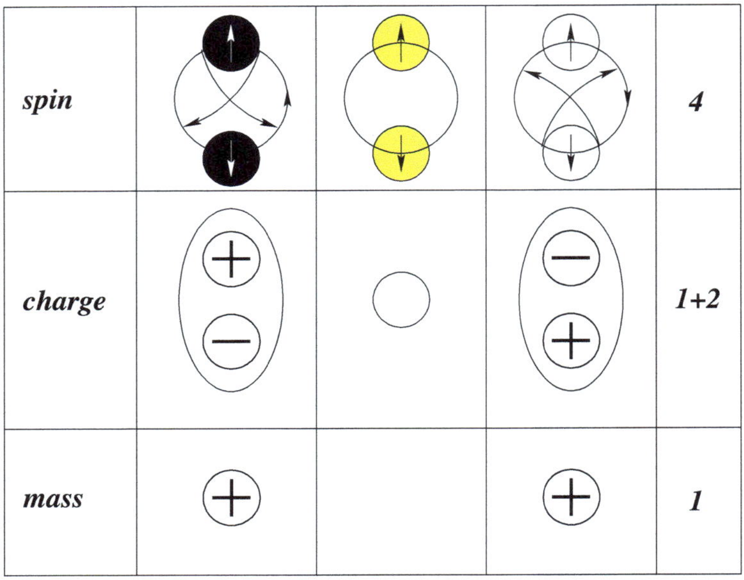

affirms a huge difference between such an environment and the impression it makes on a casual observer in tangent space. A likely image is sketched by Luminet [4]:

> ⋯ consider a room paneled with mirrors on all six surfaces (including the floor and the ceiling) ⋯ the interplay of multiple reflections will immediately cause ⋯ the impression of seeing infinitely far in every direction. Cosmic space, which is seemingly gigantic, might be lulling us with a similar illusion.

To be surrounded by a hollow spherical mirror would be even more confusing. Projective space-time that curves like a golden spiral also enhances the confusion because of self-similar symmetry and the lack of a large-scale metric. It is not inconceivable that the images of the night sky are all generated by projective distortion of local four-dimensional effects within the milky way, which itself may appear inflated in size. As cautioned by Luminet [4]:

> Treating ⋯ global aspects of space requires a mixture of advanced mathematics and subtle cosmological interpretations.

As experimental evidence of a universe, restricted in size, he [4] refers to missing long wavelengths in the cosmic microwave background and argues that the universe is simply too small to sustain such wavelengths.

Another word of caution: the so-called ‘concordance’ model adhered to by modern cosmologists considers the topology of space as strictly independent of time. A space-time model would certainly require mathematics, even more advanced, and cosmological observations, more subtle, than currently contemplated. It is safe to conclude that the size of the cosmos is completely unknown. Jumping to conclusions would be like theorizing without data; running the risk of arriving at mysterious conclusions like those of pioneer quantum mechanics.

The ultimate aim of twentieth century quantum theory was to explain the nature of matter and its interaction with radiation. The latter issue has been explored in mathematical detail as the science of spectroscopy. The success of this venture was ensured by the analysis of all interactions in terms of a simple wave theory. At all levels, from the atomic, through the molecular, to solid and plasma states, the interaction only considers the frequency of radiation and relevant electronic energy levels. The physical structure of the electrons remains unspecified.

It is only when the theory is extended to analyze the physical nature of matter in terms of quantum particles, that it tends to fall apart. As a seminal assumption it is accepted that particles of elementary matter occur in a void. The nature of the particles are, in the first instance, only known from empirical observation and mathematical details of the behaviour are modelled to fit. The derived interaction parameters appear to be delicately fine-tuned by chance, in a process that miraculously favours one out of an infinite variety of possible universes.

Interactions at the sub-atomic level, which appear to be decidedly non-classical, are controlled by probability waves that resonate with human consciousness in order to create physical matter from virtual particles in the vacuum. Apparently there is no void after all.

This, in a nutshell, is quantum field theory. It is rather awkward to describe it in the subtle detail it deserves without an exposition of probability-wave theory and the nature of virtual particles. However, as pointed out by Feynman, these considerations are too deep for mortal understanding; hence items of dogmatic belief?

The legacy of QFT is the dream of unlimited free energy, waiting to be extracted from the vacuum, actively pursued by countless amateurs. The concept is traced back to the Dirac sea that postulates the existence of an infinite number of quantum states occupied by antimatter in the vacuum. In QFT each of these states is associated with a virtual particle, mathematically described by a harmonic quantum oscillator that carries half a quantum of zero-point energy.

Hopefully the reader may find the alternative wave theory more digestible. As hinted before, it is based on the existence of an aether with physical properties. In a hypothetical Euclidean state it resembles a liquid in laminar flow. Any geometrical distortion produces turbulence, vortices, eddies and other nonlinear deformations. In a closed topology such disturbances will be of a periodic nature and, in line with special relativity, best described as four-dimensional standing waves.

All of these waves must carry the property of fermionic spin. One type of wave could be longitudinal, corresponding to penetrative neutral matter such as neutrinos. In vacuum physics, neutrino radiation is called zero-point radiation. In cosmology it is known as cosmic background radiation. Other varieties could be of a transverse nature that imparts electric charge. The simplest type of leptonic wave has a single transverse component, whereas more complicated baryonic waves have three components.

The typical lepton is an electron that carries a negative charge. The proposed baryonic structure is reminiscent of the quark model as derived from the study of symmetry groups. The proposed wave structure explains why free quarks are never observed.

The wave structure proposed for baryons is described topologically as a trefoil knot, the simplest non-trivial knot. It has a chiral structure, closely related to that of Boy's surface, but is invertible.

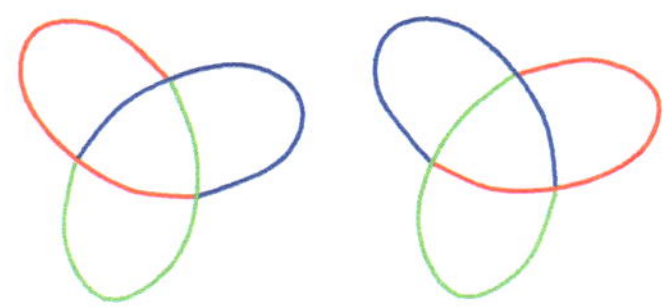

This property allows for the definition of an antibaryon as well as a baryon with inverted spin. The trefoil knot is also tricoularable, the property that may have inspired the theory of QCD.

In a projective double cover, chiral waves transform into anti-wave forms of opposite chirality as they progress through the projective involution. All waves acquire mass to a larger or lesser extent by virtue of their intrinsic nonlinear environment. When localized into a non-dispersive shape, the mass centre moves at speeds less than that of light.

The speed of light is realized only in the achiral interface between matter and antimatter. Only bosonic standing waves occur in this interface which is truly void. As the medium of communication between emitters and acceptors, these waves transmit radiant energy at the speed of light. Constrained within the interface electromagnetic radiation has only two transverse components of vibration.

All of the proposed forms of elementary wave are routinely observed as cosmic rays, including neutrinos. Apparent stellar sources are no more than secondary sources. The preponderance of these waves, which occur at a sub-critical level, are a manifestation of space-time curvature. Should their concentration exceed the critical level, that depends on the degree of curvature, more complex aggregate forms of matter are formed.

Aggregation of elementary matter does not produce atomic matter more complex than neutrons, hydrogen and helium, but this is sufficient to generate clouds that condense under gravity and collapse into stars, galaxies and black holes. The genesis of ponderable matter is briefly outlined in Appendix E. Black holes are responsible for the synthesis of heavy nuclei which are dispersed through Einstein-Rosen bridges across the vacuum interface. Excess matter is exchanged by this mechanism, resulting in cosmic equilibrium.

An unused quantity of elementary matter remains dispersed through the universe, slowly building up into interstellar clouds by combining with cosmic dust and debris. In turn, this process results in the formation of new stars and galaxies. Matter and antimatter remain effectively separated by the interface and prevented from interacting by virtue of inverted time directions. The second law of thermodynamics is balanced across the interface and the universal entropy remains at zero. There is neither heat-death nor big-bang birth and the cosmos remains in well-balanced equilibrium.

Intuitively aware of such equilibrium the search for a unification of the gravitational and electromagnetic fields became an early priority. The first unsuccessful crack by Weyl, by way of a gauge principle, later succeeded in terms of a complex phase. In this form it found application in the formulation of classical electrodynamics and the modern theory of elementary-particle physics. A convincing mathematical elucidation of gauge phenomena was provided only recently [5]. The theory developed by these authors is summarized well in their own words from the Abstract:

> Usually Maxwell's equations are invariant with respect to a gauge transformation of the potentials and one can choose freely a gauge condition. For instance, the Lorentz gauge condition yields the potential Lorentz inhomogeneous equations. It is possible to introduce a scalar field in the Maxwell equations such that the generalised Maxwell theory, expressed in terms of the potentials, automatically satisfy the Lorentz inhomogeneous wave equations, without any gauge condition. This theory of electrodynamics is no longer gauge invariant with respect to a transformation of the potentials: it is electrodynamics with broken gauge symmetry.

The essence of their theory is outlined in the Conclusions:

> It is possible to describe classical electrodynamics in the form of two biquaternion equations. This form is very useful in order to generalise electrodynamics. Generalising the Maxwell equation by introducing an extra scalar field is comparable with Maxwell's introduction of the displacement current that allowed for the derivation of the homogeneous field wave equations.

The biquaternion that they mention, is the same as an octonion, recognized before as a descriptor of a projective double cover in eight dimensions. This is independent confirmation of the proposal to relate gauge phenomena directly to projective geometry. As a potential consequence the authors [5] mention a possible 'classical theory of photon tunneling', which also resonates with the topological effects on a field in four-dimensional curved space-time.

It is worth repeating the result discussed in Chap. 4 of how projective relativity finally demonstrated the unification of the electromagnetic and gravitational fields. Gravity is a simple manifestation of space-time curvature, but electromagnetism is sensitive to a gauge factor. In projective space-time the gauge field is fixed topologically and the field equations of projective relativity hence refer to unified fields. The demonstration that the extra parameters correctly describe the electromagnetic field is achieved by transformation to tangent space, resulting in the relativistic Klein-Gordon wave equation. Transformation of the electromagnetic potentials is mediated by the golden ratio.

Most of the matter in the universe is in the plasma state. Complex inorganic matter is largely confined to planets with metallic or rocky cores, enveloped by liquid or gaseous clouds. Organic matter occurs on fewer planets and to a lesser degree in interstellar dust. The chemical interactions that result in the formation of complex forms of matter are adequately explained as the interference of electron waves.

At the end of this odyssey it is fair to ask:

"Exactly what did Max Planck discover?"

According to his own account it was that a discrete function relates the energy of a standing wave to its frequency. In symbols $E = (nh)\nu$, where h is a constant and n an integer. The constant has dimensions of energy×time, called *action*. Black-body radiation therefore teaches that the action of a standing wave is an integral multiple of an elementary unit.

A standing wave has a fixed length and oscillates between displacements of A and $-A$, as shown. Each vibrating unit of wavelength λ stores an amount of energy $E = hc/\lambda = h\nu$. Over one period of oscillation, $\tau = 1/\nu$, the unit of action, $E\tau = h$. The total action of a standing wave, stretching over n wavelengths, is nh and the energy, $E = nh\nu$, often ascribed to n photons.

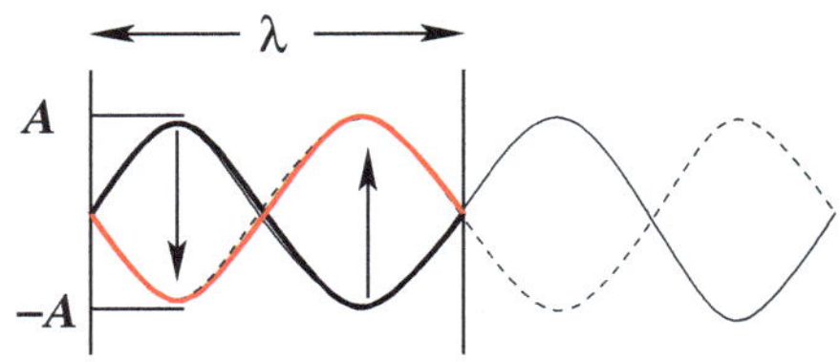

In four-dimensional space-time one unit of action represents 2π units of spin. This means that what Planck discovered was the spin of standing waves, which indeed is the property that distinguishes non-classical waves from their three-dimensional counterparts.

The momentous discovery of electron diffraction should have put the nature of all matter in proper perspective, but by then Bohr's quantum theory of the atom had already entrenched the particle model of the electron. Schrödinger's linear wave equation caused even more confusion, which can only be undone by an equation in four-dimensional nonlinear spinors.

At this point the Copenhagen story is history. What remains are electromagnetic waves together with three types of matter wave corresponding to neutrinos, electrons and protons. The former propagate through the vacuum in the projective interface at the speed of light, not impeded by any aether drag. A word of caution is in order at this point. The concept of speed cannot be generalized into Minkowski space-time and has operational meaning only in a three-dimensional world. Propagation in the vacuum interface is best described in terms of standing waves. Nonlinear matter waves are in the aether with de Broglie wavelength of h/mv and variable velocity $v < c$, which depends on the local environment. The wavelength adapts to increasing nonlinearity until the wave breaks up into individual solitons that move through the aether without getting entangled, by virtue of their spherical spin.

As the concentration of elementary matter waves reaches super-critical levels, atomic matter is formed by the following interactions:

$$p^+ + e^- + \nu \rightarrow n$$
$$p^+ + e^- \rightarrow \text{H}$$
$$2p^+ + 2n \rightarrow \alpha^{2+}$$
$$\alpha^{2+} + 2e^- \rightarrow \text{He}$$

Under conditions of extreme gravity the chemical elements are synthesized by the fusion of α particles [6]. As the nuclides pour out from a synthetic black hole into more rarefied intergalactic space, they become less stable and many of them disintegrate completely. The 264 stable isotopes of 81 elements that occur in the solar system are readily demonstrated to be adapted to the space-time curvature characteristic of a golden logarithmic spiral. Not only nuclide stability but the variation of all chemical properties are faithfully modelled by golden-spiral optimization and elementary number theory.

The prospect of formulating a new theory for chemistry with minimal reference to traditional quantum mechanics appears to be bright. As a theory of chemistry, the quantum gamble has not paid off and the other sciences should take note.

Apart from the quantum of action the concepts that feature most prominently in this work are normed division algebras and projective geometry. In concluding the argument it is appropriate to mention the close relationship that exists between these concepts.

In terms of the complex pair $u = a + ib$ and $v = c + id$ the quaternion is defined as

$$q = a + ib + jc + kd \ ; \ q = a_0 + a_ie_i$$

with the rule of composition: $i^2 = j^2 = k^2 = ijk = -1$, summarized graphically by the cyclic triplet shown below.

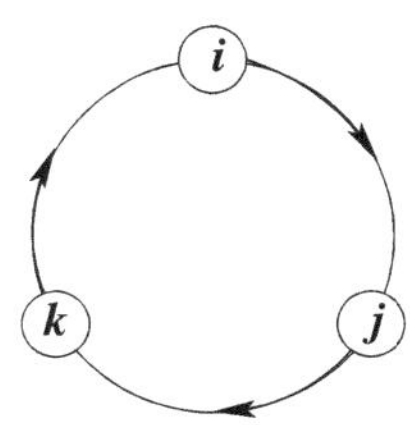

Multiplication of two elements going clockwise produces the next one, e.g. $ij = k$. The product going anticlockwise is $ji = -k$.

By extending the process of dimension doubling to a pair of quaternions an 8-dimensional non-associative division algebra of octonions, of the form

$a_0 + \sum_{i=1}^{7} a_i e_i$ for all a_i real, is produced. The rule of multiplication is summarized by the following *Fano plane*, which consists of 7 points and 7 directed lines, counting the central circle as a line.

Each pair of points lies on a unique line. Each line contains three points and each of these triplets has a cyclic ordering shown by the arrows. In cyclic order $e_i e_j = e_k$; $e_j e_i = -e_k$, $e_i^2 + e_j^2 + e_k^2 = -1$, a precise copy of the two quaternions within the octonion, $e_1 - e_2 - e_4$ and $e_3 - e_6 - e_5$.

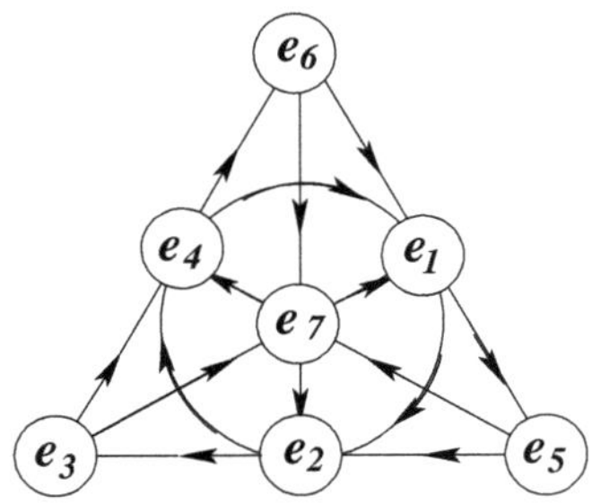

All other properties of division algebras are satisfied. The superimposed pair of quaternions represent a left- and right-handed pair of spinors. The Fano plane also represents the double cover of the projective plane over a two-element field of integers [7], which becomes important in the definition of antimatter. It also represents the projective plane with the fewest points and lines. Any pair of lines intersect in only one point.

References

1. N. Tesla, *New York Herald Tribune*. Sept. 11 (1932)
2. G.K. Francis, *A Topological Picturebook* (Springer, New York, 1987)
3. J. Stillwell, *Geometry of Surfaces* (Springer, New York, 1992)
4. J.-P. Luminet, *The Shape and Topology of the Universe*. arXiv:0802.2236v1[astro-ph] 15 Feb 2008
5. K.J. van Vlaenderen, A. Waser, Generalization of classical electrodynamics to admit a scalar field and longitudinal waves. Hadronic J. **24**(2001), 609–628
6. J.C.A. Boeyens, D.C. Levendis, *Number Theory and the Periodicity of Matter*. Springer.com (2008)
7. J.H. Conway, D.A. Smith, *On Quaternions and Octonions* (A K Peters, Natic MA, 2003)

Appendix A
Quaternion Rotations

The same problem that led Hamilton to the discovery of quaternions also puzzled the rocket scientists behind the Apollo space project on running into a gimbal (or robot) lock. In the end, the lunar mission was also saved by quaternions.

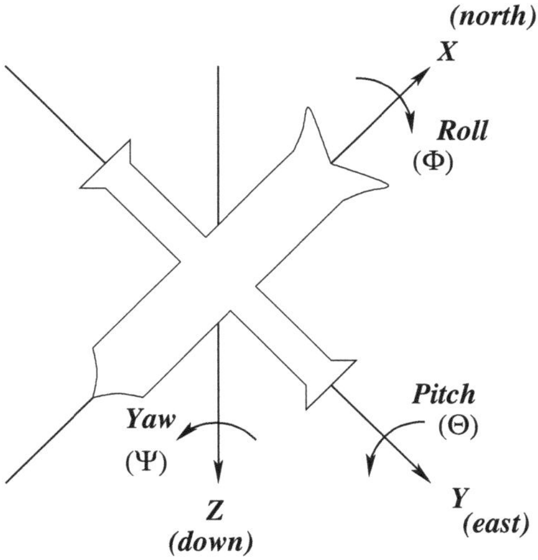

Gimbal lock occurs when a mechanical device locks up, or fails to function, because one degree of freedom, among the three exes of rotation, is lost. It is a real danger in an automobile factory that relies on the use of robotic arms. Any system that uses Euler angles of rotation about three orthogonal axes to control orientational change, may have problems with gimbal lock. It occurs during MRI (Magnetic-Resonance Imaging) and CAT (Computed Axial Tomography) scans, also known as CT. The reason for this is that rotations on Euler axes are independently evaluated in a set order, as illustrated by Hamilton's coloured cubes. Gimbal lock occurs on

J.C.A. Boeyens, *The Quantum Gamble*, DOI 10.1007/978-3-319-41621-2

a $\pi/2$ rotation about the second axis, which leaves two axes lined up in the same direction. The essential problem is that Euler rotations, known in practice as pitch, roll and yaw, are not a mathematically complete description of the orientation of an object, as Hamilton discovered.

In the case of an aircraft, or a space capsule, the orientation sensors encode rotation from the 'inertial frame' to the 'body frame'. On rotating the capsule to the point where the Y-axis of the body frame coincides with the inertial Z-axis, the confusion between pitch and yaw amounts to a singularity, loss of the sense of orientation and inability to control the spacecraft.

This problem, and of all quoted examples, can be solved completely by using a system with four degrees of freedom to represent rotations. Quaternions are ideally suited for this purpose. Any combination of yaw-pitch-roll can be reproduced by a single quaternion rotation, as demonstrated by Hamilton's cubes, but not by Eulerian rotations.

Appendix B
Separation of the Variables

The procedure that immediately suggests itself for solution of the equation

$$\Box^2 V = \left[\frac{\partial^2}{\partial x_0^2} + \frac{\partial^2}{\partial x_1^2} + \frac{\partial^2}{\partial x_2^2} + \frac{\partial^2}{\partial x_3^2}\right] V = 0$$

is based on separation of the variables by assuming that the potential is adequately defined as a product function. It is assumed that the entangled function

$$V(x_\nu) = V_0(x_0) \cdot V_1(x_1) \cdot V_2(x_2) \cdot V_3(x_3).$$

If the product function can be decomposed the four simpler one-dimensional problems can be solved separately and the results recombined. The procedure is demonstrated for the two-dimensional case that implies

$$V(x, y) = X(x) \cdot Y(y)$$

as in

$$\frac{\partial^2 V}{\partial x^2} + \frac{\partial^2 V}{\partial y^2} = 0\,.$$

The product function is differentiated twice,

$$\frac{\partial^2 V}{\partial x^2} = \frac{d^2 X}{dx^2} \cdot Y$$

$$\frac{\partial^2 V}{\partial y^2} = \frac{d^2 Y}{dy^2} \cdot Y$$

to give

$$\frac{d^2 X}{dx^2} \cdot Y + \frac{d^2 Y}{dy^2} \cdot X = 0$$

J.C.A. Boeyens, *The Quantum Gamble*, DOI 10.1007/978-3-319-41621-2

Divide by $V = X \cdot Y$:

$$\frac{1}{X}\frac{d^2X}{dx^2} + \frac{1}{Y}\frac{d^2Y}{dy^2} = 0$$

The first term is independent of y and the second is independent of x, as a function of only y. Each term can therefore be separately equated to a constant, e.g.

$$\frac{d^2X}{dx^2} = KX = k_x^2 X$$

This equation is readily solved by assuming K to be a squared quantity,[1] which implies

$$V = e^{(k_x x + k_y y)}\ , \quad k_x^2 + k_y^2 = 0$$

This condition requires that either $k_y = ik_x$ or that $k_x = k_y = 0$, *i.e.*

$$V = X \cdot Y = c_x e^{\pm kx} \cdot c_y e^{\pm iky} = ce^{\pm k(x+iy)}$$

or V = constant. Separation of space and time variables results in $V(x,t) = V(x)e^{-i\omega t}$ where $\omega = 2\pi\nu$ is an angular frequency.

1

Let $$\frac{d^2X}{dx^2} = k^2X\,, \quad \frac{d}{dx} = D$$

hence, $$(D^2 + k^2)X = 0$$

This expression is a special case of the more general family of equations

$$(D - p_1)(D - p_2)X = 0$$

i.e. $$D^2X - (p_1 + p_2)DX + p_1p_2X = 0$$

This equation reduces to the special case for $p_1 = -p_2$, which implies that $p_1^2 = -k^2\,,\ \ p_{1,2} = \pm ik$. For each term

$$(D - p_1)X = 0\,,\ \ i.e.\ \frac{dX}{dx} = p_1X \ \text{ or } \ \frac{dX}{X} = p_1 dx$$

Integration gives $$\ln X = p_1 x + const \ \ (= \ln A)$$

Thus $$\ln\frac{X}{A} = p_1 x = \pm ik$$

which is $$X = A\exp(ikx) + B\exp(-ikx).$$

The first solution defines a complex exponent in which the variables are mathematically entangled, which geometrically represents a rotation. In the second solution the variables are separated, with the geometrical meaning of two orthogonal vibrations. Vital information is irretrievably lost by this separating of the variables.

Returning to the four-dimensional equation it is noted that solution as a product function requires that $k_0^2 + k_1^2 + k_2^2 + k_3^2 = 0$. This condition is completely satisfied only by a function in which the four variables are entangled on an equivalent basis in the form $z = a_0 + a_1e_1 + a_2e_2 + a_3e_3$ where the e_i are generalizations of $\sqrt{-1}$. The most general solution, in this hypercomplex form is known as a *quaternion*. Quaternions are known to describe both Lorentz transformation and four-dimensional action, known as *spin* in quantum theory. This property is lost in three dimensions on separating out the time variable. Schrödinger's wave equation approximates four-dimensional action by a superposition of angular momentum and one-dimensional vibration. On further separation of the variables, as practised in quantum chemistry, angular momentum also disappears as the theory collapses into classical mechanics. A spin parameter which is added empirically in the same style is responsible for the dubious perception of non-local quantum effects.

Appendix C
Broken Gauge Symmetry

In an attempt to unify the electromagnetic and gravitational fields the idea of gauge transformation was first proposed by Herman Weyl as a space-time dependent change of scale on displacement from point x^μ to $x^\mu + dx^\mu$. Weyl could demonstrate that such a change of gauge generates the potentials of the electromagnetic field in four-dimensional space-time.

An obvious objection to Weyl's theory is that taking a clock through a closed loop in four-dimensional space-time, must change the speed of the clock and an atom carried around a closed path in an electromagnetic field must therefore radiate at a different wavelength on reaching the end of the loop. This is not confirmed experimentally.

It was demonstrated by Schrödinger that Weyl's prescription, with a complex gauge factor, predicted the correct quantized energy levels for the hydrogen electron, enabling London to reformulate the gauge transformation as a phase change

$$\Psi \rightarrow \Psi' = \Psi \exp(iq\alpha) \tag{C.1}$$

of the wave function of a charged particle. As in the original proposal the phase factor may have a globally fixed value, independent of position, in which case (C.1) is known as a *global* gauge transformation. This applies in the pseudo-Euclidean Minkowski space of special relativity and the transformation has the special property that the derivative of the field

$$\partial_\mu \Psi = e^{i\alpha} \partial_\mu \Psi$$

transforms like the field itself. If the complex field Ψ describes an electron, a charge q is associated with its wave field of charge density $\rho = |\Psi|^2$, which is clearly invariant under gauge transformation (C.1). This invariance implies, by Noether's theorem, the global conservation of the total charge, and shows that the overall phase change is not measurable.

J.C.A. Boeyens, *The Quantum Gamble*, DOI 10.1007/978-3-319-41621-2

The situation is entirely different within the theory of general relativity, which is based on a curved manifold rather than flat space with a globally fixed coordinate system. Each point now has its own coordinate system and hence its own (local) gauge factor. By doing away with the rigid coordinate system the gauge factor necessarily becomes an arbitrary function of the coordinates, $\alpha(\mu)$. Since the phase has no real physical significance, it may be defined locally by an arbitrary rotation, at every space-time point of the manifold, without changing the physical situation. This stipulation may seem to rule out local charge conservation, unless there is some compensating field that restores the invariance under local phase transformation.

In the case of an electron wave function the required gauge field must be the electromagnetic field. To find the correct form of this compensating field we look for the transformed wave function

$$\Psi' = \Psi e^{i\alpha(x,t)}$$

that satisfies Schrödinger's equation:

$$i\hbar\frac{\partial\Psi}{\partial t} = -\frac{\hbar^2}{2m}\left(\frac{\partial^2\Psi}{\partial x^2}\right) \tag{C.2}$$

The function $\Psi'(x,t)$ clearly does not satisfy the equation, yielding:

$$i\hbar\frac{\partial\Psi}{\partial t} = -\frac{\hbar^2}{2m}\left[\frac{\partial^2\Psi}{\partial x^2} + 2i\frac{\partial\alpha}{\partial x}\frac{\partial\Psi}{\partial x} - \Psi\frac{\partial^2\alpha}{\partial x^2} - \Psi\left(\frac{\partial\alpha}{\partial x}\right)^2\right] + \hbar\Psi\frac{\partial\alpha}{\partial t}$$

This expression is conveniently abbreviated in vector notation to read:

$$\begin{aligned} i\hbar\frac{\partial\Psi}{\partial t} &= -\frac{\hbar^2}{2m}\left[(\nabla + i\nabla\alpha)^2 - \nabla\cdot\nabla\alpha - \frac{2m}{\hbar}\frac{\partial\alpha}{\partial t}\right]\Psi \\ &\equiv -\frac{\hbar^2}{2m}\left[(\nabla + i\boldsymbol{A})^2 - \boldsymbol{\nabla}\times\boldsymbol{A} - V\right]\Psi \end{aligned} \tag{C.3}$$

The form of equation (C.2) is therefore recovered by defining the vector $\boldsymbol{A} = \partial\alpha/\partial x$ and the scalar

$$V = \frac{2m}{\hbar}\cdot\frac{\partial\alpha}{\partial t}.$$

Equation (C.3) is recognized as the well-known Pauli equation that describes the motion of an electron in an electromagnetic field. $\boldsymbol{A}$ and V are respectively known as the vector and scalar potentials of the electromagnetic field. The quantity $\boldsymbol{\nabla}\times\boldsymbol{A} = \text{curl}\boldsymbol{A} = \boldsymbol{B}$ represents the magnetic field strength. In order to leave the electromagnetic field invariant, the gauge field itself must transform like $A'_\mu \rightarrow A_\mu(x) + \partial_\mu\alpha(x)$.

The complex phase which is fundamental to gauge theory and in particlephysics is defined in terms of symmetry groups without consideration of its physical meaning, which emerges most clearly in its characterization of the quantum wave functions. Whereas phase relationships between point particles are hard to imagine, they appear naturally in wave structures. With respect to electrons and other chemical entities a wave model in terms of complex wave functions is therefore the most satisfactory physical model. The complex phase represents the fundamental attribute of non-classical systems and the major difference between classical particles and quantum waves.

In a sense, the terminology of global and local gauge invariance may be somewhat unfortunate. We distinguish between Riemannian underlying space-time and local Minkowski tangent space. The invariant gauge factor in tangent space is commonly said to define global gauge invariance whereas local gauge invariance is associated with the curved underlying space.

To summarize, the principle of local gauge invariance in a curved Riemannian manifold leads to the appearance of compensating fields. The electromagnetic field is the compensating field of local phase transformation and the gravitational field of local Lorentz transformations. Such compensating fields cannot be scalars, but are vector or tensor fields whose associated rest mass vanishes. In particle physics the interaction between an electron and the electromagnetic field is said to be mediated by a massless photon, which leads to the local phase invariance. The four-vector $\boldsymbol{A}$ describes the electromagnetic field, or the wave functions of the photon.

The difference between global and local gauge invariance explains the apparent differences in behaviour of radiation and matter in tangent (field-free) space and underlying curved space. The conservation of electric charge in tangent space becomes subject to the generation of an electromagnetic field in curved space. This means that the stability of an electron depends on its interaction with the environment.

The behaviour of massive objects differs in an analogous way in tangent space and the underlying curved space. In tangent space the motion of a test object obeys Newton's law of inertia, but in curved space it becomes subject to gravitational effects.

Gauge invariance, as formulated here, represents the elusive link between general relativity and quantum theory.

The central idea of particle physics is inspired by this successful description of an electron in terms of the electromagnetic gauge field. Having recognized the appearance of strong and weak interactions in atomic nuclei it was argued that these could also serve to characterize the corresponding elementary particles by specifying the appropriate interaction as a gauge field. Whereas Schrödinger's equation correctly describes electronic interaction, the starting point in the analysis of strong and weak interaction is to ensure that the Lagrangian function, which describes the interaction of these particle wave functions, remains invariant under the symmetry transformations that reflect known conservation laws.

The next important step is to identify asymmetric solutions of the symmetrical Lagrangian. This procedure corresponds in principle to the modification of global gauge invariance by the recognition of a local gauge field, as before. The process has become known as *spontaneous symmetry breaking* of gauge theories. The most important application in the present context is the broken symmetry that generates the Higgs field, characterized by a massive boson.

This result is important within elementary-particle physics as a mathematical procedure to simulate the effect of space-time curvature, which is known to produce matter. In this sense it is neither unexpected nor mysterious. However, the common interpretation of the effect as a transition between two symmetry states of the vacuum has no observational support. More obviously, the high-symmetry Lagrangian is the correct formulation in hypothetical Euclidean space, whereas the 'broken' symmetry represents physically real curved space-time. Another way of distinguishing between the two symmetry states is in terms of linear and nonlinear formulations.

It is emphasized that the Higgs mechanism does not refer to two physically realizable vacuum states in that pseudo-Euclidean Minkowski space-time is no more than a local tangent approximation to the curved manifold of general relativity. It has no independent existence.

'Symmetry breaking' only has mathematical meaning and it makes no sense to associate it with a phase transition between two possible vacuum states. A Higgs field is one mathematical manifestation of space-time curvature.

Appendix D
Projective Geometry

As in algebra, any exception or special case in the theory of geometry, is interpreted to indicate a suspect axiom. Euclid's axiom of parallel lines defines such a special case. The problem is that a pair of parallel lines differs from any other pair of straight lines that invariably intersect in a point. A possible remedy is to postulate two parallel lines to intersect at infinity. However, by the same assumption these lines should then also intersect at minus infinity, i.e. at two points in total. To avoid this exception the only way out is to consider the two points at $\pm\infty$ as a single point. Geometrically this is achieved by adding an extra point at infinity to each real line—the geometrical equivalent of adding the number zero between positive and negative numbers.

This construction is not possible in the Euclidean plane. The so-called projective plane that contains the lines of interest curves into the third dimension and to keep the lines from coalescing this plane has a half-twist or an *involution*. The Möbius band clarifies the situation. It is noted that the extra point at infinity can be placed anywhere along the central line in the surface.

To envisage the topology of the closed projective plane it is necessary to start at a point in the edge of the Möbius band and stitch the diverging edges together. It is soon evident that this operation cannot continue to completion in three-dimensional space without cutting through the band. The projective plane is said to be *embedded* in four dimensions.

J.C.A. Boeyens, *The Quantum Gamble*, DOI 10.1007/978-3-319-41621-2

Whereas most theorems of geometry are concerned with the concepts distance, angle and congruence, a smaller number of others are only concerned with the incidence of points and straight lines. This distinction differentiates between the common *metrical* properties of geometry and those, which are independent of measurement, and which reflect the characteristics of what became known as *projective geometry*. This property has the same effect as Weyl's proposed gauge (phase) invariance that features in the definition of the electromagnetic field.

A fundamental operation in projective geometry is the mapping of points on a line (l) onto another (l') from a central point P, as shown in the figure below.

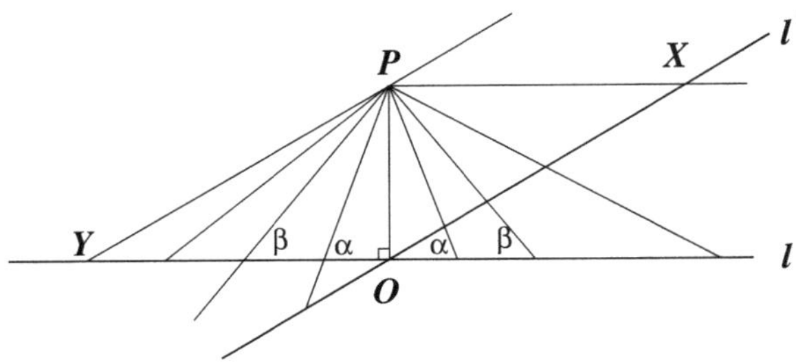

Each point on l projects to a unique point on l' and each possible line through P connects two, uniquely defined (homologous) points on l and l', with one exception. The perpendicular on OP through P fails to intersect l and hence the point X has no homologue on l. By the same reasoning the point Y on l has no homologue on l'. This creates an awkward dilemma because there is nothing special about the points X and Y. Points on either side of X are, for instance, successfully mapped to l. The problem lies with the lines XP and YP.

Starting from PO, which is perpendicular to l, the angles of intersection, α, β, $\gamma \ldots$, decrease monotonically until XP is reached with a zero angle of intersection. By continued rotation of the ray X/P the angle increases again, through Y, to a maximum at OP. This situation is reminiscent of the problem with the number system before the discovery of zero, which was solved by the addition of a fictitious integer, 0. The geometrical problem is likewise solved by the addition of ideal points at infinity on the lines l and l'. This discovery is as momentous for geometry as the discovery of zero was for algebra. It means that two lines, traditionally considered as parallel, can now be defined to intersect at an ideal point. The concepts of parallel and endless lines therefore fall away. There is no distinction between the ideal point and the regular points on a line. Any regular point may therefore also be interpreted as a point at infinity.

The mapping shown in the figure also illustrates the procedure commonly used in making perspective drawings.

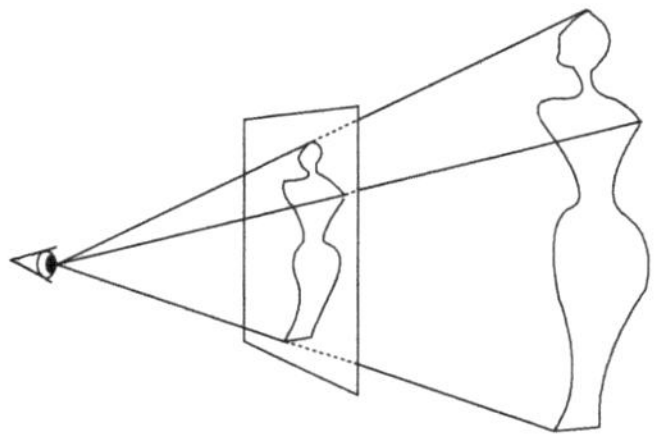

In this case the central point is the eye of the artist. The line l represents the scene of interest and l' the canvas. Each radial line registers as a point at P and the plane of projection appears as a line, such as l. All parallel lines share the same ideal point and all ideal points at infinity appear as the ideal line at infinity, or *horizon*, which terminates the plane of projection. In three dimensions all parallel lines contain the same line at infinity and finally, all points and lines at infinity form a single plane at infinity.

The foregoing is interpreted to mean that the projective model of space is closed by a single surface that corresponds to the ideal plane at infinity. In Euclidean geometry this plane appears curved. By assuming that the structure of the cosmos is subject to mathematical analysis and that the mathematics applies without exception, it is a logical necessity that the geometry of space-time be projective.

Points on the projective line are arranged in such a way that their order is cyclic. This implies that all conics, including the straight line, are equivalent in projective geometry. The Euclidean relations of parallelism, betweenness, order and congruence therefore have no meaning. It can be stated that, without exception, any two lines in the projective plane intersect in a point, any two points in projective space define a line and any two planes intersect in a line.

Radial lines in projection look like points and radial planes look like lines when viewed edge-on, which means that radial dimensions are lost. Because of this, physical models formulated in five-dimensional Euclidean (or affine) space are reproduced in four-dimensional projective space without the awkward concept of compacted dimensions. This correspondence defines the principle of *duality* which asserts that any definition or theorem in projective geometry remains valid on interchanging the words *point* and *line*.

The important final conclusion is that n-dimensional Euclidean space is comprehensively mapped into $(n-1)$-dimensional projective space. A fifth dimension is therefore not required to define four-dimensional curved space-time projectively.

The projective plane has some remarkable special properties. Imperfectly expressed in words, because of the extra points at infinity, any regular point in n-dimensional projective space is described by $n+1$ *homogeneous* coordinates. The consequence of this is that, unlike any other topological space, projective space has no *metric*. By implication, there is no clear distinction between the concepts of point, line and plane and the homogeneous coordinates remain valid when multiplied by an arbitrary factor, $x^{\alpha} = kx^{\alpha}$. This property is not unlike Weyl's conjecture that on displacement in curved space-time the magnitude of a vector may change. Interpreted as a multiplicative *phase* factor, this conjecture eventually resulted in the fundamental concept of *gauge invariance*, at the root of modern elementary-particle theory.

As for a Möbius band, the projective plane is one-sided and said to be non-orientable, because of the involution. The Möbius surface is obtained as a slice through a projective plane. It is of zero thickness and closed through an involution. A double cover is obtained by glueing a strip of finite thickness on to the whole length of the one-sided Möbius surface. It differs from the primary surface as it only closes on itself after two inversions. The original surface now occurs as a two-dimensional interface between two three-dimensional bands, related to each other by inversion symmetry. A short cross section of the resulting construct is shown on the left in the diagram.

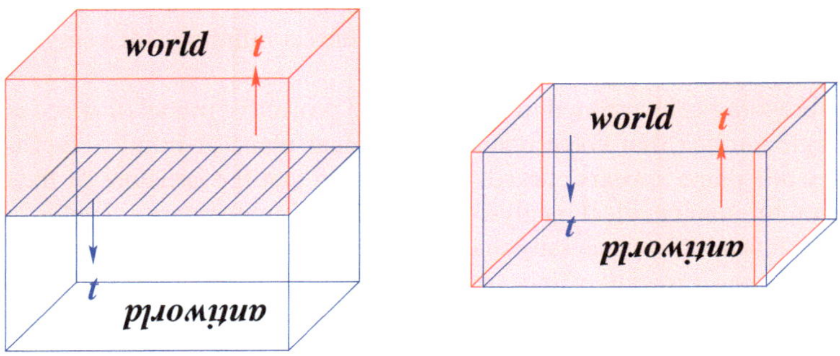

A doubly-covered projective plane is obtained by stitching together the open edges of the doubly-covered Möbius surface, and that will be proposed as a model of the universe. In the same way that the Möbius surface cannot be constructed in a two-dimensional plane, the projective world is embedded in four dimensions, with a three-dimensional interface, interpreted as the physical vacuum, separating antipodal regions of the four-dimensional double cover.

In the composite drawing shown on the right-hand side of the diagram, the antipodal world and antiworld appear to occupy the same space, but since the direction of time is also inverted by the involution, they are separated by the interface in space-time.

Following a straight line in the surface for one complete revolution, one ends up at the opposite side of the interface, but still in the same surface; the same property that defines the spin function. The resulting antipodal points are of opposite chirality. The geometry of the double cover can be shown to be correctly described in terms of 8-dimensional hypercomplex (octonionic) algebra.

Being one-sided, all points in a projective surface, as on a Möbius band, are equivalent. The hair on such a surface can all be brushed in the same direction without developing a crown. Hence it has no unique central point. There is no other closed space like this and it will be argued to define the most likely topology for the four-dimensional universal space-time vacuum.

Appendix E
Nucleogenesis

It is known from elementary number theory [1] that the 264 stable isotopes of the natural elements occur as 11 periodic groups of 24 that contain the periodic table as a subset. On replotting this distribution as a function of atomic number the hemlines that define the periodic function of 24 are no longer parallel to one

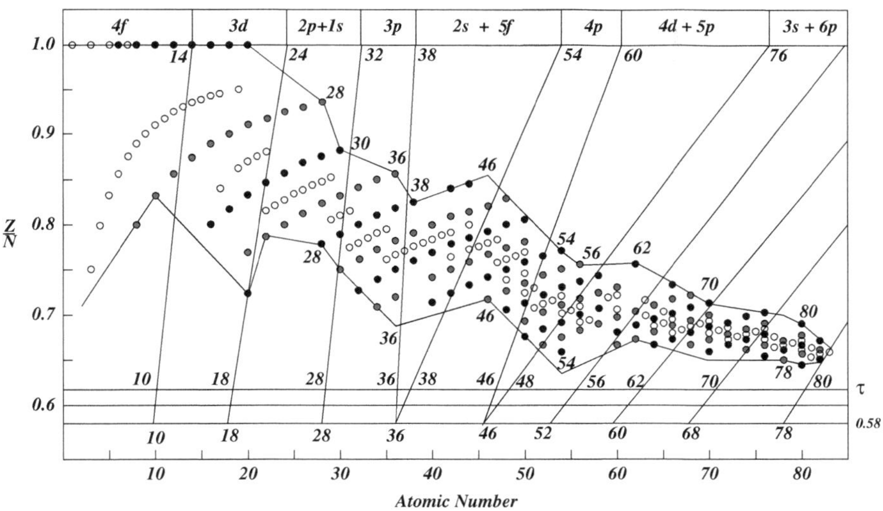

another and lines of constant Z/N intersect the hemlines at variable values of Z. Taking a cue from the convergence, established to occur at $Z/N = \tau$, it is noticed that this line intersects the hemlines at values of Z, familiar from the structure of the experimentally established Periodic Table of the Elements.

On extending these lines to $Z/N = 0.58$ intersections now occur at values of Z/N, characteristic of the periodicity predicted by Schrödinger's one-electron equation for hydrogen. The points of intersection at $Z/N = 1$ define a periodic

J.C.A. Boeyens, *The Quantum Gamble*, DOI 10.1007/978-3-319-41621-2

table with inverted hydrogen energy levels. The inflection points of an empirical stability profile invariably correspond to the closure of electronic energy levels.

Crossing of energy levels is a known effect, resulting from isotropic compression of an atom. The symmetry of s, p, d, f spectroscopic states decreases in this order and isotropic compression that enhances the symmetry, affects the relative energies in the reverse order. From this observation it is inferred that the periodicity at $Z/N = 1$ represents a highly compressed state of atomic matter. On a cosmic scale the Schrödinger hydrogen atom represents an empty universe, at zero pressure. The other extreme is conjectured to occur in massive stars and black holes. Under these conditions fusion of α-particles leads to the formation of nuclides with $N/Z = 1$.

On release into intergalactic space intra-atomic rearrangements render all, but 300, of these nuclei unstable against β-decay. In time, α-decay further reduces the number of stable nuclides to 264, as observed in the $Z/N = \tau$ periodic system.

An intriguing possibility suggested by the convergence of Z/N to the golden ratio, is that the arrangement of elementary proton and neutron waves in the nucleus may be related to the botanical phenomenon known as Fibonacci phyllotaxis, beautifully illustrated in the structure of a cornflower.

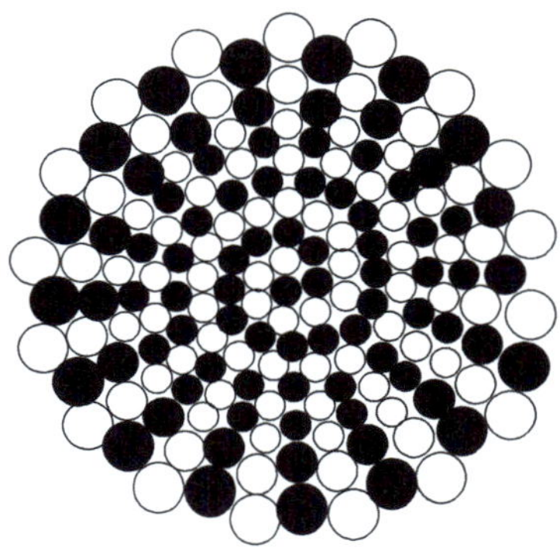

The number of florets that spiral conspicuously in opposite sense, always corresponds to successive numbers in a Fibonacci sequence. The ratio between successive Fibonacci numbers, like Z/N for atomic nuclei, converges to the golden ratio. In a three-dimensional analogue, proton and neutron waves may be envisaged to spiral in complementary fashion, as shown by the schematic diagram in cross section. Now assume that the wavelengths of proton and neutron waves are in the ratio of τ, such that Z protons in a unit sphere match τN neutrons. Noting that for all stable nuclides $Z/N > \tau$, *i.e.* $Z > \tau N$. Hence, the envisaged spherical golden-ratio phyllotaxis must always leave an excess of $x = Z - \tau N$ protons in the nuclear surface, to facilitate electrostatic interaction with the electron cloud. The only exception is ^{3}He, an isotope with relative abundance less than one part in ten thousand. The convergence of Z/N to τ for stable nuclides is now seen as convergence to the most stable arrangement of nucleons in strong interaction.

For $Z = \tau N$ the positively charged surface disappears and the electron cloud dissipates. The convergence of nuclear compositions to a ratio of $Z/N = \tau$, which may, in a sense, be interpreted as an approach to maximum stability, is never realized, being in conflict with an atom's ability to stabilize the extranuclear charge cloud. Elements with $Z > 83$ decay by α emission because of this effect.

Reference

1. J.C.A. Boeyens, D.C. Levendis, *Number Theory and the Periodicity of Matter*. Springer.com (2008)

MIX
Papier aus verantwortungsvollen Quellen
Paper from responsible sources
FSC® C105338
FSC
www.fsc.org

If you have any concerns about our products,
you can contact us on
ProductSafety@springernature.com

In case Publisher is established outside the EU,
the EU authorized representative is:
Springer Nature Customer Service Center GmbH
Europaplatz 3, 69115 Heidelberg, Germany

Printed by Libri Plureos GmbH
in Hamburg, Germany